UNDERSTANDING
BIOCHEMICAL PATHWAYS
A PATTERN-RECOGNITION APPROACH

FIRST EDITION

By Carol A. Wilkins

Bassim Hamadeh, CEO and Publisher
Michael Simpson, Vice President of Acquisitions and Sales
Jamie Giganti, Senior Managing Editor
Miguel Macias, Senior Graphic Designer
John Remington, Senior Field Acquisitions Editor
Monika Dziamka, Project Editor
Brian Fahey, Licensing Specialist
Berenice Quirino, Associate Production Editor
Chris Snipes, Interior Designer

Copyright © 2018 by Cognella, Inc. All rights reserved. No part of this publication may be reprinted, reproduced, transmitted, or utilized in any form or by any electronic, mechanical, or other means, now known or hereafter invented, including photocopying, microfilming, and recording, or in any information retrieval system without the written permission of Cognella, Inc. For inquiries regarding permissions, translations, foreign rights, audio rights, and any other forms of reproduction, please contact the Cognella Licensing Department at rights@cognella.com.

Trademark Notice: Product or corporate names may be trademarks or registered trademarks, and are used only for identification and explanation without intent to infringe.

Cover image copyright © 2013 by iStockphoto LP / isak55.

Printed in the United States of America

ISBN: 978-1-5165-0997-3 (pbk) / 978-1-5165-0998-0 (br)

TABLE OF CONTENTS

Dedication	iv
Acknowledgments	v
Introduction and Purpose of Textbook	vi
Chapter 1: Oxidation State Patterns	1
Chapter 2: Metabolism Overview and Glycolysis	31
Chapter 3: Mitochondrion Overview, the Pyruvate Dehydrogenase Complex, and the Tricarboxylic Acid Cycle	65
Chapter 4: The Electron Transport Chain	87
Chapter 5: Gluconeogenesis and the Pentose Phosphate Pathway	99
Chapter 6: β-oxidation of Fatty Acids and Ketone Body Synthesis	131
Chapter 7: Fatty Acid Synthesis	151

DEDICATION

To my son Brett, my biggest fan!

To the Michigan State University College of Human Medicine Advanced Baccalaureate Learning Experience (ABLE) students, Classes of 1998 through 2011, for constantly encouraging me to publish my "recipe" to benefit others.

ACKNOWLEDGMENTS

I have been blessed to grow up in a loving and supportive family, as well as marry into an equally loving and supportive one. Most importantly I would like to thank my husband David and my son Brett; my parents, Jim and Jean Mindock; my brothers Jim, Joe, and Daniel and their families; my in-laws, Charlie and Joan, Kim, Mark, and Steve and their families. I am very grateful for all you have done for me.

I have had numerous wonderful teachers and mentors throughout my schooling and career. I want to acknowledge several of them who have influenced me in how I teach biochemistry. The first is Michigan State University (MSU) emeritus professor John (Jack) F. Holland. He had a wonderful way of summarizing metabolic pathways using his "Pawn Shop Model" of metabolism. He also taught his students the rules for naming enzymes, which I also incorporated into my teaching of metabolism. Another is MSU emeritus professor, Richard L. Anderson. He emphasized that one should be able to answer four questions regarding each metabolic pathway: What? Why? Where? How? I also incorporated his key questions in teaching my courses, and I have expanded on his idea to six main questions. Answering these six key questions about each pathway provides foundational understanding for the pathways and metabolism, in general.

I would like to extend my thanks and appreciation to three of my mentors, MSU professors John L. Wang and John J. LaPres, as well as Dr. Donald J. Sefcik at Marian University for their continual support and advice. They have been influential in helping me in my career and in how I teach.

I would also like to thank my students at Michigan State University, past and present, who are a constant source of inspiration.

Finally, I want to thank my two cats, Houdini and Nikita. They have been wonderful companions to me and my family for many years. Through my stories about them, they have also helped many of my students remember the regulation of fatty acid oxidation as CAT I (Houdini—the regulated CAT) and CAT II (Nikita—the unregulated CAT).

INTRODUCTION AND PURPOSE OF TEXTBOOK

Understanding Biochemical Pathways: A Pattern-Recognition Approach is meant to be used in conjunction with any other complete biochemistry textbook that is typically required for any level of undergraduate, graduate, or medical school biochemistry course. This is not a stand-alone biochemistry textbook. The purpose of this textbook is to provide the reader with a pattern-recognition approach to understanding metabolic processes (with the emphasis on human metabolic processes), and it focuses on *specific pathways* of carbohydrate and lipid metabolism that illustrate how to apply this approach. This text presents a "basic recipe" of metabolism to illustrate the general sequence of reactions that are carried out in biochemical pathways to move from one oxidation state of carbon to another. The goal is to give the reader the ability to look at a reaction and determine the type of reaction (based on the differences in reactants and products), identify the type of enzyme that would catalyze the reaction, and then name the enzyme based on enzyme naming rules. However, this text does *not* convey or imply the actual enzymatic mechanism by which the reaction is carried out. This text also gives a specific set of questions for the reader to answer regarding any given pathway to assist in understanding the "big picture" of the metabolic pathway and a means for understanding how various metabolic pathways are regulated and integrated.

This text therefore presumes the reader is learning (or has learned) certain basic biochemical structures and concepts. These structures include basic structures of amino acids, lipids, nucleotides, and vitamins. The reader should have knowledge of protein structures, enzyme kinetics, and types of enzyme regulation. The reader should also have an understanding of Gibbs free energy concepts (ΔG), as this plays an important role in understanding how the concentrations of metabolites can influence the direction in which a reaction proceeds.

This approach is not as simple or easy as one might assume from the standpoint that there is still much to learn about each pathway. This approach will reduce the amount of rote memorization, which many students feel is the only way to learn metabolic pathways, and facilitate better long-term understanding of metabolism. Once one recognizes various patterns in a pathway, key structures, and enzyme naming rules, one can learn the "list of reactions" of a pathway. The reason is now one has a picture of the metabolic pathway that is based on key structures. One also understands what has to happen to get from the beginning molecules to the end products of the pathway and the types of reactions and enzymes needed based on the list.

In using this approach to understanding biochemical pathways, the intent is for the reader to be able to build further upon this knowledge to the extent required by his or her own interest and/or the biochemistry course in which he or she is enrolled. Biochemistry is a fascinating and complex subject, in which there are many facets that are not fully understood or are being newly discovered. One's understanding of biochemistry will always be advancing. The pattern-recognition approach presented in this text will hopefully serve as a foundation upon which one can build this additional knowledge.

CHAPTER 1

OXIDATION STATE PATTERNS

OBJECTIVES

1. Define basic organic chemistry terms and apply them to biological reactions.
    a. Define the terms: oxidation, reduction, hydration, dehydration, protonate, deprotonate, hydrogen atom, proton, hydride ion, alpha (α) carbon, beta (β) carbon, and omega (ω) carbon.
2. Recognize the structures of functional groups of carbons and their corresponding prefixes/suffixes used in scientific naming of molecules.
    a. Recognize the following functional groups for carbons: alkane, alkene, alcohol [primary (1°), secondary (2°), and tertiary (3°)], aldehyde, ketone, and carboxylic acid.
    b. Recognize the corresponding suffixes used in naming molecules that will indicate the presence of the various functional groups in biological molecules.
3. Recognize the relationship between corresponding α-amino acids and α-keto acids.
4. Recognize the common reactions used in biological pathways.
    a. Compare any two of the various functional groups listed above to distinguish which of the pair is more oxidized, which is more reduced, or if they are both at the same oxidation level.
    b. Identify the oxidation states of carbons that *cannot* be further oxidized.
5. Recognize various reaction types and how the enzymes that carry them out are named; especially regarding dehydrogenases, reductases, kinases, synthases, and synthetases.
6. Diagram the basic reaction patterns needed to oxidize or reduce carbons on a biological molecule.
    a. As oxidation and reduction reactions are always coupled, identify the most likely donor or carrier of hydrogen atoms for the reaction (i.e., $NAD^+/NADH$, $NADP^+/NADPH$, or $FAD/FADH_2$).
7. Explain the significance of having a keto group in the α- or β-position relative to a carboxylic acid.

INTRODUCTION

The first part of this chapter will review some basic concepts and terminology from organic chemistry regarding how to number carbons and recognize key functional groups, as well as nomenclature tips for naming of molecules. Key organic chemical reactions that are carried out in metabolic pathways will also be reviewed. In the second part of this chapter, a "basic recipe of metabolism" will be presented that will form the basic pattern of reaction sequences that commonly occur in metabolic pathways. An overview of key points in how metabolic pathways work to make energy and the basic rules for naming enzymes are also covered in this chapter.

BASIC CONCEPTS AND TERMINOLOGY

A. Carbon numbering and nomenclature tips

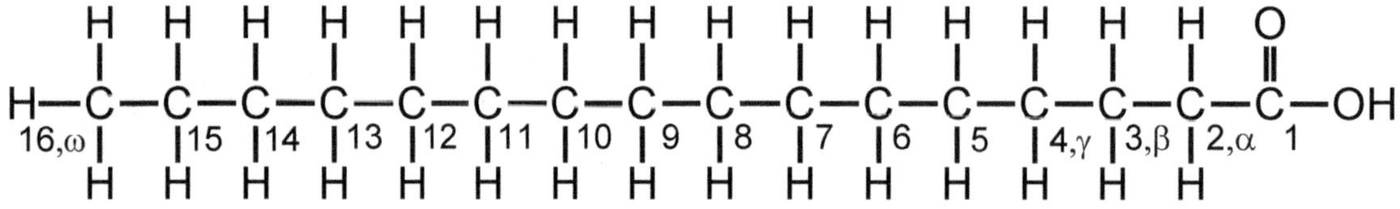

Figure 1.1: Palmitic acid ($C_{16:0}$), a fatty acid.

There are two general methods of numbering carbons. For the scientific method of numbering carbons, the carbon with the highest oxidation state is generally designated carbon 1. If the carbon with the highest oxidation state is not an end carbon of the molecule, the end carbon nearest the carbon with the highest oxidation state is designate carbon 1 (i.e., the ketoses). The remaining carbons are numbered sequentially from carbon 1. The hierarchy of the carbon oxidation states will be covered later in this chapter. For the fatty acid molecule drawn in **figure 1.1**, the carboxylic acid carbon is the highest oxidation state of carbon in this molecule and thus is designated carbon 1. The remaining carbons are numbered sequentially to the methyl carbon at the other end, which is carbon 16 for the fatty acid drawn.

Another carbon designation system is often used in reference to biological molecules, which involves labeling carbons using Greek letters. This nomenclature is reserved for molecules that have a carboxylic acid. So the carbon of the carboxylic acid is *not* designated with a letter. The carbon *next* to the carboxylic acid carbon is designated "alpha." For the fatty acid in **figure 1.1**, it is carbon 2 that is the alpha (α) carbon. Carbon 3 is the beta (β) carbon. This will be important to note when discussing β-oxidation of fatty acids, in which carbon 3 (the beta carbon) is the carbon oxidized. Carbon 4 is the gamma (γ) carbon, and so forth. Using this nomenclature for fatty acids, however, the end methyl carbon of a fatty acid, regardless of its length, is the omega (ω) carbon. For the fatty acid shown in **figure 1.1**, carbon 16 is the omega (ω) carbon. This terminology is important in understanding the significance of the "omega fatty acids," which refers to the relationship of the double bonds in the fatty acid relative to the omega end of the fatty acid molecule.

For acid groups that are in the protonated state, the name of the molecule includes "-ic acid" as in palmitic acid for the fatty acid drawn in **figure 1.1**. Other examples of this nomenclature include acetic acid, pyruvic acid, and lactic acid. When the acid group is in the deprotonated state, for instance the carboxylate ion COO^-, (the typical state for most acid groups at physiological pH), it is indicated in the molecular name as "-ate" (i.e., palmitate, acetate, pyruvate, lactate). Thus, the molecular name indicates whether the acid group is in the protonated or deprotonated form.

The "-yl" ending, which is commonly seen in molecular naming, means that a functional group of a particular molecule is now attached to another molecule. For example, when a fatty acid is attached to another molecule like coenzyme A, the molecule is called "fatty ac*yl* CoA," or often more simply "acyl CoA." In another example, when three fatty acids are attached to glycerol, the new molecule is called a "triac*yl*glycerol." Another important example of this "-yl" nomenclature is used in distinguishing the attachment of acetic acid to coenzyme A. Acetic acid (or "acetate" for the deprotonated form) is the shortest fatty acid and is commonly found in biological systems. This fatty acid is two carbons long and contains a methyl group attached to the carboxylic acid group. When acetate is attached to coenzyme A, the molecule is called "acet*yl* CoA" and is a very important intermediate of metabolic pathways.

B. Key reaction definitions

Oxidation, also known as de<u>hydrogen</u>ation, is the loss of electrons. For organic reactions, oxidation is the loss of hydrogens. Notice the whole word "hydrogen" is in the name "dehydrogenation". Hydrogens carry the electrons. As biological systems are not made of "metal wires," they need a carrier for the electrons of these reactions, which are usually hydrogens (though there are exceptions). For the typical oxidation reactions in the catabolism of carbohydrates, fatty acids, and amino acids, these reactions involve the loss of *two* electrons—or loss of *two hydrogens*.

Reduction, also known as <u>hydrogen</u>ation, is the gain of electrons. For organic reactions, reduction is the gain of hydrogens. Again, these reactions generally gain two electrons or hydrogens at a time. Oxidation and reduction reactions are always coupled. Biological systems cannot just let go of electrons and release them, because biological systems are not metal wires. So if a molecule is oxidized, then another molecule must be reduced (or vice versa). Therefore, these reactions need coenzymes (derived from vitamins) to carry the hydrogens, which include NAD^+, $NADP^+$, FMN, and FAD. These coenzymes are involved in oxidation-reduction reactions because they are carrying the hydrogens to or from an organic molecule. There are many examples that will be reviewed in metabolic reactions. The key point for now is to remember that oxidation-reduction reactions are always coupled and are typically referred to as "redox" reactions.

Now contrast oxidation (dehydrogenation) and reduction (hydrogenation) reactions with dehydration and hydration reactions. Dehydration and hydration reactions involve the loss and gain of an entire water molecule, which are *not* redox reactions. Another set of reactions to differentiate are protonation and deprotonation reactions (see **figure 1.2**). These reactions involve the gain (<u>proton</u>ation) or loss (de<u>proton</u>ation) of protons (H^+). These reactions are acid-base chemistry, *not* oxidation-reduction reactions, because a proton does not have an electron on it. Thus, it is very important to be specific in distinguishing a proton, a hydrogen atom, and a hydride ion and the types of reactions in which they are involved.

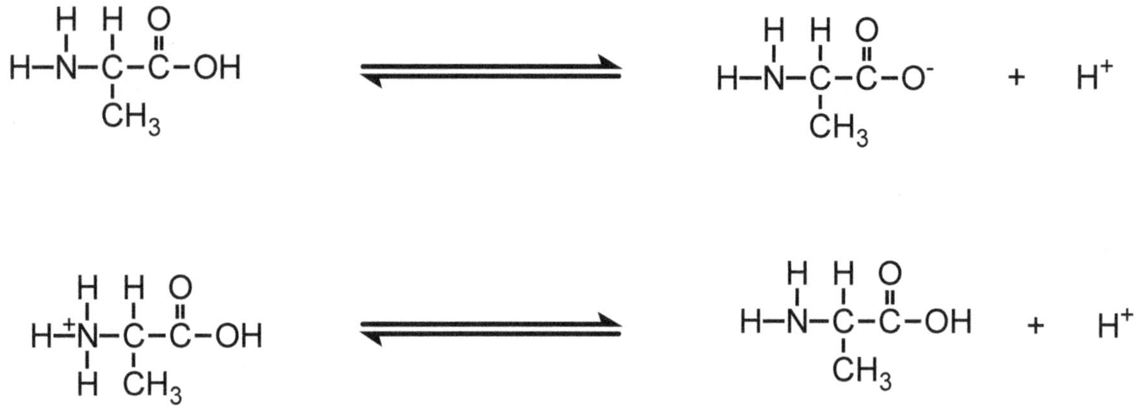

Figure 1.2: Protonation/deprotonation reactions. These are acid-base reactions, not oxidation-reduction reactions.

C. Hydrogen terminology

As indicated in **table 1.1**, a hydrogen atom has one proton and one electron. For organic reactions in most biological systems, hydrogen atoms serve as the carrier of electrons. So to lose electrons from a molecule, the molecule loses hydrogens. If the molecule is gaining electrons, the molecule picks up hydrogens.

Table 1.1: Hydrogen terminology

1. Hydrogen atom	H = 1 proton and **1** electron
2. Hydride ion	H^- = 1 proton and **2** electrons
3. Hydrogen ion (a.k.a. proton)	H^+ = 1 proton and **NO** electrons

A hydride ion is a hydrogen atom carrying *two* electrons (H^-). Molecules like NAD^+ and $NADP^+$ have room for only one hydrogen atom. While the organic molecule typically loses two hydrogens (i.e., two electrons), NAD^+ and $NADP^+$ then pick up a hydride atom carrying both electrons, and the reaction involves a "leftover proton," as will be explained further in the next section.

A proton, then, is a hydrogen atom that has one proton and *no* electrons; hence the name "proton." H^+ is also referred to as a "hydrogen ion." In this text, the term "proton" (H^+) will be used to prevent confusion with hydride ion. Again, this is why acid-base reactions (deprotonation-protonation) are different than oxidation-reduction reactions. Acid-base reactions move protons (not electrons), while oxidation-reduction reactions move electrons usually using either hydrogen or hydride atoms.

D. Reactions to consider

Vitamins provide the basis for the coenzyme derivatives needed for many reactions. In this chapter the focus will be on the coenzymes needed for various redox reactions (see **figure 1.3**). If a molecule is being oxidized, a coenzyme needs to be reduced. If a molecule is being reduced, a coenzyme needs to be oxidized. Throughout metabolic pathways, many examples of these reaction types will be reviewed.

Nicotinamide adenine dinucleotide (NAD$^+$) can only accept one hydrogen atom, while the molecule being oxidized is typically losing two hydrogen atoms. Thus, both electrons need to end up on one hydrogen atom, which leaves a leftover proton (NADH + H$^+$). This description does not indicate or imply anything about the actual enzyme mechanisms by which these reactions occur. Rather, the reader should note

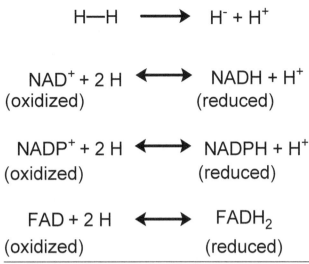

Figure 1.3: Electron transfer of hydrogens, and coenzymes involved in oxidation-reduction reactions.

the establishment of the "pattern" of noting how many hydrogens are lost or gained from an organic molecule, as well as the type of electron carrier accepted or lost by the coenzyme (i.e., either two actual hydrogen atoms or a hydride ion plus a leftover proton). Nicotinamide adenine dinucleotide phosphate (NADP$^+$) carries out the same reaction as NAD$^+$, meaning that it can only accept a hydride ion (leaving a leftover proton), because NADP$^+$ is the same molecule except it has a phosphate "tag" as part of its structure.

Another reason to note that electrons can be stripped from a hydrogen atom to produce a proton leads to the concept of proton "pumping" in the electron transport chain. The basic concept of the electron transport chain is to remove the electron from a hydrogen atom. In the electron transport chain some of the coenzymes only accept electrons because they do contain metal atoms. This results in the production of a proton. The movement of protons from the mitochondrial matrix to the intermembrane space ultimately produces a proton gradient.

The vitamin riboflavin is converted into two coenzyme derivatives: flavin adenine dinucleotide (FAD) and flavin mononucleotide (FMN). Both have the same functional part of the molecule, and they have room for both hydrogen atoms that are lost from a molecule. So FAD and FMN get reduced to FADH$_2$ and FMNH$_2$, respectively.

NADH and FADH$_2$ are commonly referred to as "reducing power" because they are the carriers of hydrogens (electrons) to the electron transport chain. NADH and FADH$_2$ are "reducing power" because they donate their electrons to another molecule to "reduce it" in the electron transport chain. The electron transport chain is the pathway that produces the majority of adenosine triphosphate (ATP), the major form of chemical energy for a cell, from the catabolism of biological molecules. NADPH, on the other hand, is the form of reducing power used for synthetic reactions, overall. NADPH is not used as an electron donor to the electron transport chain.

E. Primary, secondary, and tertiary designations of alcohols

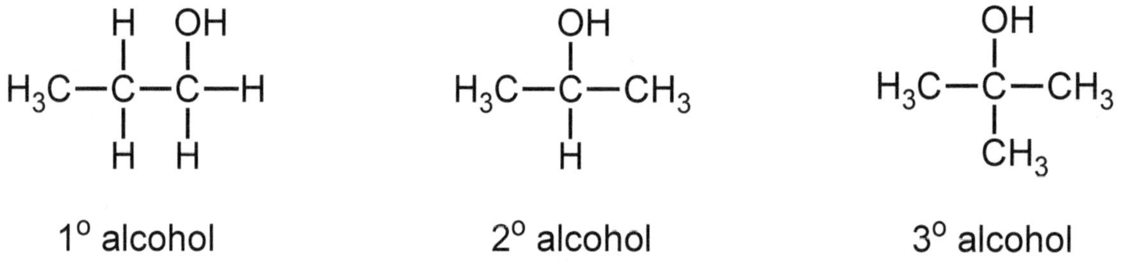

$$\underset{1°\ \text{alcohol}}{\text{H}_3\text{C}-\overset{\overset{\text{H}}{|}}{\underset{\underset{\text{H}}{|}}{\text{C}}}-\overset{\overset{\text{OH}}{|}}{\underset{\underset{\text{H}}{|}}{\text{C}}}-\text{H}} \qquad \underset{2°\ \text{alcohol}}{\text{H}_3\text{C}-\overset{\overset{\text{OH}}{|}}{\underset{\underset{\text{H}}{|}}{\text{C}}}-\text{CH}_3} \qquad \underset{3°\ \text{alcohol}}{\text{H}_3\text{C}-\overset{\overset{\text{OH}}{|}}{\underset{\underset{\text{CH}_3}{|}}{\text{C}}}-\text{CH}_3}$$

Figure 1.4: Primary, secondary, and tertiary alcohols.

Another set of basic chemistry terms to review is the designation of primary, secondary, and tertiary alcohols (see **figure 1.4**). In this case the designation is based on the carbon with the hydroxy group (-OH) on it. If this carbon is attached to only one other carbon, it is a primary alcohol. If the hydroxy carbon is attached to two other carbons, it is a secondary alcohol. If the hydroxy carbon is attached to three other carbons, it is a tertiary alcohol. The key to note here is that tertiary alcohols *cannot* be further oxidized. For example, in the tricarboxylic acid (TCA) cycle, the reason citrate must be isomerized to isocitrate is because citrate contains a tertiary alcohol.

F. Relationship between α-amino acids and α-keto acids

Recall the alpha (α) carbon is the carbon next to a carboxylic acid. In **figure 1.5** the α-keto acids are drawn on the left. These molecules must have a keto group next to a carboxylic acid. Ketones must be "in the middle" of a molecule, where the carbonyl carbon (carbon with a double bond to oxygen) is attached to two other carbons. Recall that a carbon atom can only make four bonds. The top left molecule of **figure 1.5** is pyruvate, an α-keto acid. Its name does not indicate much about its structure, other than it has a carboxylic acid group that is in the deprotonated state. However, pyruvate is such a key molecule in metabolism that it is absolutely one of the key structures the reader should know (i.e., be able to draw it, as well as recognize it).

Now "replace" the α-keto groups for the molecules on the left of **figure 1.5** with an amino group and a hydrogen atom, which results in the molecules drawn on the right. These are now α-amino acids, which have attached to the α-carbon: a carboxylic acid, an amino group, a hydrogen, and an R group. The amino acid drawn at the top right is alanine. This is the simplest type of reaction to add or remove the nitrogen (as an amino group) in the synthesis or catabolism of amino acids. Simply "remove" the amino group to produce the corresponding α-keto acid, or create a new amino acid by putting an amino group on an α-keto acid. Thus, this is the relationship between corresponding α-keto acids and α-amino acids. Again, this description is not intended to imply anything about the enzyme mechanism by which this done, only noting the structural pattern between α-keto acids and α-amino acids. The reaction that accomplishes this interchange is called transamination, which is carried out by enzymes called *aminotransferases*. In some older texts, the enzymes are called transaminases.

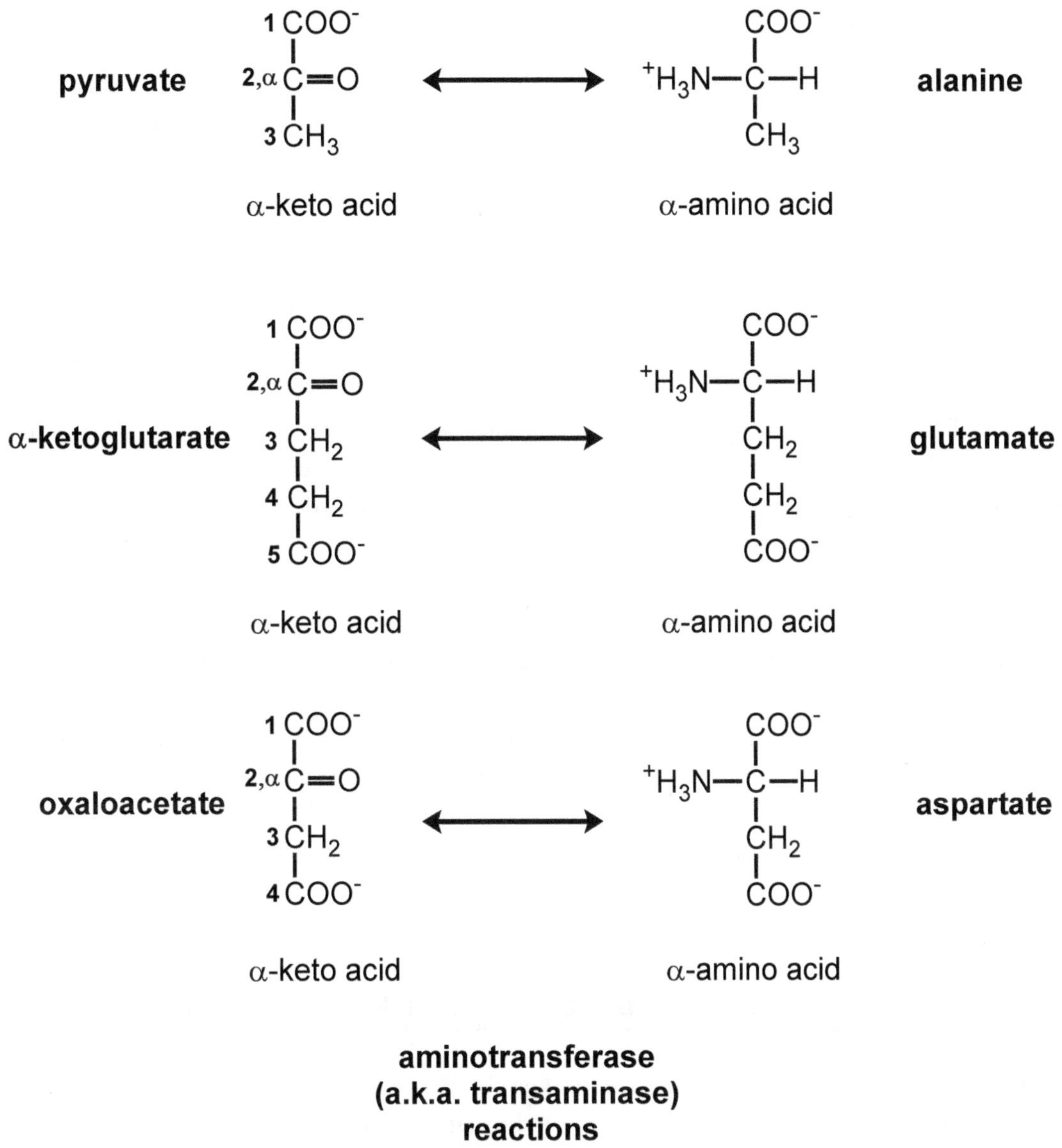

pyruvate $\quad$ α-keto acid $\longleftrightarrow$ α-amino acid $\quad$ alanine

α-ketoglutarate $\quad$ α-keto acid $\longleftrightarrow$ α-amino acid $\quad$ glutamate

oxaloacetate $\quad$ α-keto acid $\longleftrightarrow$ α-amino acid $\quad$ aspartate

aminotransferase (a.k.a. transaminase) reactions

Figure 1.5: Relationship between α-keto acids and α-amino acids.

A couple of other corresponding α-keto acids and α-amino acids are shown in **figure 1.5** for review. For another nomenclature hint: anytime the prefix "glut" is in a molecular name—think five carbons. For the middle left molecule drawn in **figure 1.5**, the carbons are numbered 1 through 5, with a keto group on carbon 2 (the α-carbon) producing an α-keto acid. This α-keto acid is called α-keto<u>glut</u>arate. If the α-keto group is replaced with an amino group and a hydrogen, the corresponding α-amino acid, <u>glut</u>amate, is produced (shown in **figure 1.5**).

OXIDATION STATE PATTERNS

The bottom set of molecules drawn in **figure 1.5** show the α-keto acid oxaloacetate and its corresponding α-amino acid, aspartate. Again, the exchange of a keto functionality for an amino group and hydrogen atom differentiates the two molecules. Both α-ketoglutarate and oxaloacetate are tricarboxylic acid (TCA) cycle intermediates. Thus, knowing the relationships between α-amino acids and α-keto acids will provide basic understanding of connections to amino acid synthesis and breakdown pathways.

Even if the reader is not required to draw structures on assignments or examinations, he or she must be able to recognize key structures. Learning structural patterns will help the reader differentiate structures without necessarily being able to draw entire molecules (and requires less rote memorization of structures). For instance, glutamate and aspartate are the two amino acids that have carboxylic acid functionalities as part of their R groups. Remembering that "glut" means five carbons, one needs to recognize the basic structure of an amino acid to determine the R group portion. If the R group ends in a carboxylic acid functional group and has a total of five carbons (including the α-carboxylic acid group carbon), the amino acid must be glutamate. If the R group ends in a carboxylic acid functional group but is not five carbons, then it must be the "other one"—aspartate (which has four carbons total).

Whether or not the reader is required to draw structures, wherever possible let the name of the molecule indicate the actual structure of the molecule. First learn key structures, which will be indicated throughout this text. Then, by learning patterns of structures as well as how molecules are named, recognition or drawing of other derivatives of those key structures will become more straightforward. For example, if one can draw glutamate as its five-carbon structure, simply note that aspartate is basically "missing" one CH_2 group (rather than memorize these two entire structures as separate entities).

OXIDATION STATES FLOWCHART (BASIC RECIPE OF METABOLISM)

The following section will cover the oxidation states flowchart, or what the author has termed the "basic recipe of metabolism." IMPORTANT NOTE: In the discussion of the reactions of this oxidation states flowchart, *nowhere* are the actual enzymatic mechanisms being implied or described. This basic recipe of metabolism simply lays out the pattern of organic reactions that are carried out in humans (and many other organisms), the typical coenzymes needed, and the relationships of the types of reactions used by the body to get carbons to the various oxidation states.

The reader must first be able to recognize the differences between the various functional groups: an alkane, an alkene, an alcohol, a keto group, an aldehyde, and a carboxylic acid group. These are all various oxidation states of carbon and are indicated in **figure 1.6**. **Figure 1.6** also indicates the common suffixes used in naming conventions to identify various functional groups as components of molecules. This figure indicates the oxidation states of the various functional groups in relation to one another, which is organized from fully reduced to fully oxidized molecules. Functionalities at the same oxidation level are indicated. Full oxidation of a three-carbon alkane to three molecules of CO_2 would indicate complete oxidation of the all the carbons. Breathing off of these three molecules of CO_2 would mean that the molecule has been completely removed from the body.

Many molecules have more than one functional group, or even more than one of the same functional group. So when one is trying to determine what type of reaction is happening to go from one molecule to the next—focus on what is *different* between the molecules. If the starting and ending molecules have the same number of each atom, rearranged, they have been isomerized. If the starting molecule does not have a phosphate group and the product does, the molecule has been phosphorylated. If the phosphate group has been removed, a dephosphorylation reaction has occurred. If the only difference between the starting and ending molecules is the number of hydrogens, it is an oxidation-reduction reaction. The metabolic pathways contain many examples of these types of reactions and will be reviewed in the context of the oxidation states flowchart (the basic recipe of metabolism), which is covered in the next section.

The oxidation states flowchart, shown in **figure 1.7**, is the key figure for recognizing patterns of reactions in metabolism. This flowchart of reactions is what the author refers to as the "basic recipe of metabolism." The oxidation states flowchart, as stated, does not refer to actual mechanisms. Rather, the flowchart indicates the set of commonly repeated reactions that occur, often in sequence, in numerous metabolic pathways. These sets of reactions are most often seen in the pathways of catabolism or synthesis of monosaccharides, fatty acids, and amino acids. In the following chapters examples of how to recognize these reactions individually and/or in sequence will be demonstrated using common metabolic pathways covered in biochemistry. This will hopefully provide the reader with a basic foundation for learning additional pathways and building subsequent knowledge regarding the complexity and integration biochemical pathways.

The orientation of the functional groups drawn in the oxidation flowchart is organized with the most reduced molecule at the top of **figure 1.7**. The most oxidized molecule is at the bottom of the figure. As one moves down the page, the molecules become more oxidized relative to one another. As one moves up the page, the molecules become more reduced. Molecules drawn at the same level horizontally across the page are at equivalent oxidation states (i.e., the types of reactions needed to convert them are hydrations/dehydrations or isomerizations, *not* redox reactions).

At the top of the oxidation states flowchart (**figure 1.7**) is drawn an alkane, which is the most reduced state of carbon. An alkane is basically a carbon making four single bonds (the maximum it can make) to either carbons or hydrogens only. Thus, the three carbons in the alkane drawn are all at the most reduced oxidation state. The ultimate goal in catabolic biological pathways is to oxidize carbons to carboxylic acids, which can be clipped off and transported in the blood back to the lungs to be exhaled from the body, while each oxidation step produces "reducing power" in the form of NADH or $FADH_2$.

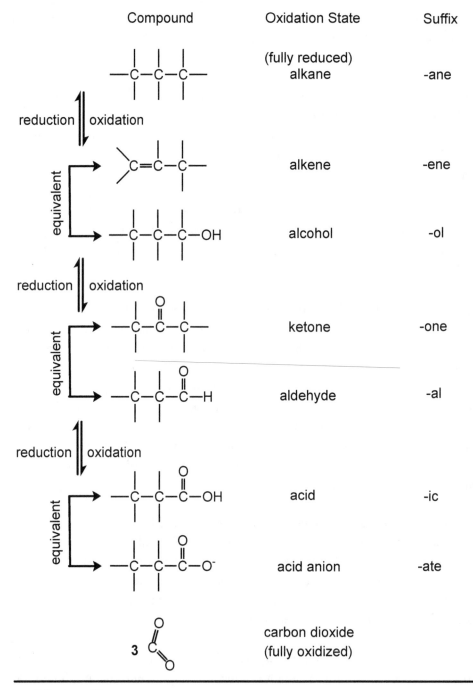

Figure 1.6: Nomenclature of oxidation states and suffixes.

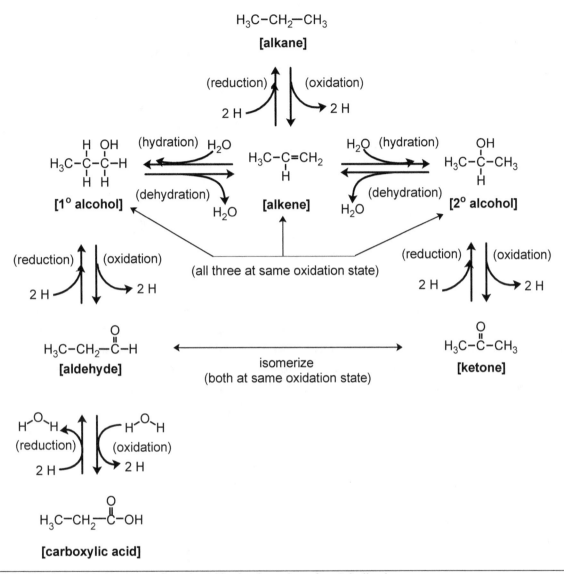

Figure 1.7: Oxidation states flowchart (the basic recipe of metabolism).

The reactions in the oxidation states flowchart will now be separated out to review them individually or in smaller sets. In each reaction covered, the substrates and products will be drawn to match the oxidation states flowchart. The first reaction is the oxidation of the alkane to the alkene, shown in **figure 1.8**.

The alkane drawn has three carbons and eight hydrogens. The alkane is converted to an alkene, which has a double bond between two carbon atoms (general formula: $R_2C = CR_2$). The alkene still has three carbons, but only six hydrogens. In this reaction the alkane has lost two hydrogens to become an alkene. Therefore, this is an oxidation reaction. So the alkene is now *oxidized* relative to the alkane. However, as previously mentioned, molecules in biological systems do not just "lose" hydrogens (i.e., electrons)—the hydrogens must go somewhere. The first general rule for coenzyme use in redox reactions: Anytime an alkane is oxidized to an alkene, FAD is used as the coenzyme. FAD is reduced to FADH$_2$. Recall that oxidation and reduction reactions are *always* coupled. Note that the indication of "oxidization" next to the down arrow of the

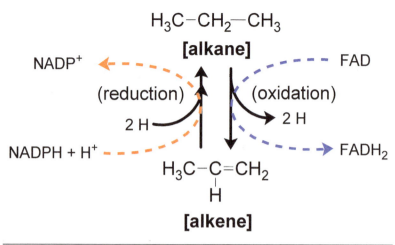

Figure 1.8: The oxidation of an alkane to alkene (and reverse reaction).

reaction (**figure 1.8**) is relative to what is happening to the organic molecule (i.e., the alkane to alkene comparison). Again, a coenzyme must always be used to accept the hydrogens (i.e., be reduced). This chart is *not* implying the actual enzymatic mechanism that takes place for the transfer of the hydrogens from the alkane to FAD. Just note the pattern and that the coenzyme for enzymes oxidizing alkanes to alkenes is FAD, which gets reduced to FADH$_2$. FAD is also used as a coenzyme for other types of redox reactions carried out by certain enzymes (i.e., the pyruvate dehydrogenase complex and the α-ketoglutarate dehydrogenase complex).

Since the other oxidation states of carbon have oxygens, oxygens will need to be added to the molecule to get the carbons to these other oxidation states. Water molecules will be used to add these oxygens. One must further note that biochemical pathways cannot go directly from an alkane to an alcohol. The alkane must be oxidized to the alkene first. Once the alkene is produced, then the alcohol functionality can be formed.

Now consider the reactions that convert the alkene to an alcohol, shown in **figure 1.9**. An alcohol is a molecule with a hydroxy group (–OH). Thus, to convert the alkene to an alcohol requires the addition of an entire molecule of water. In the hydration reactions drawn in **figure 1.9**, an entire water molecule is added across the double bond, with the placement of an –OH on one carbon and the remaining hydrogen on the other carbon. Where the –OH group is placed depends on whether a primary alcohol or secondary alcohol is produced, which was defined previously. If the hydroxy group is placed on the end carbon, as drawn on the left in **figure 1.9**, a primary (1°) alcohol is produced. If the hydroxy group is placed on the middle carbon, as drawn on the right in **figure 1.9**, a secondary (2°) alcohol is produced. Tertiary (3°) alcohols can also be formed (not shown), but tertiary alcohols cannot be further oxidized, as previously mentioned. Note that the alcohols and alkene are drawn at the same level horizontally, meaning they are all at the same oxidation state.

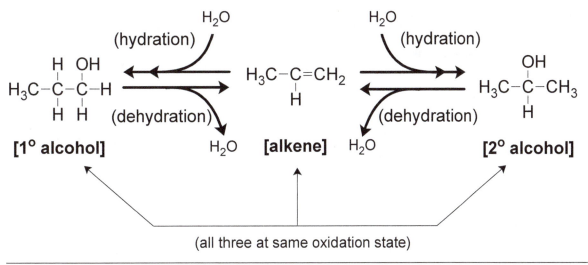

Figure 1.9: The interconversion of alkenes and alcohols.

Now consider the secondary alcohol, which can be further oxidized. In the reaction drawn in **figure 1.10**, simply count the hydrogens. Notice that two hydrogens from the secondary alcohol are lost (one from the hydroxy group and one from the center carbon—again, not implying the actual mechanism), forming a carbonyl group (also known as a keto group, or ketone). A C = O is a carbonyl carbon. There are many functional groups that have a carbonyl group (i.e., aldehydes and carboxylic acids), but a keto group is "just" a carbonyl group (C = O) with the carbon bound to two other carbons.

Since the only difference between the secondary alcohol and the ketone drawn in **figure 1.10** is the loss of two hydrogens, this is an oxidation reaction. These two hydrogens must be added to a coenzyme. In this case the coenzyme is NAD$^+$ (oxidized form), which is reduced to NADH + H$^+$. Remember there is always a leftover proton, as NAD$^+$ picks up a hydride ion carrying both electrons. In fact, NAD$^+$ is the coenzyme for all of the remaining oxidation reactions going down the flowchart. FAD is only used for the alkane to alkene reaction.

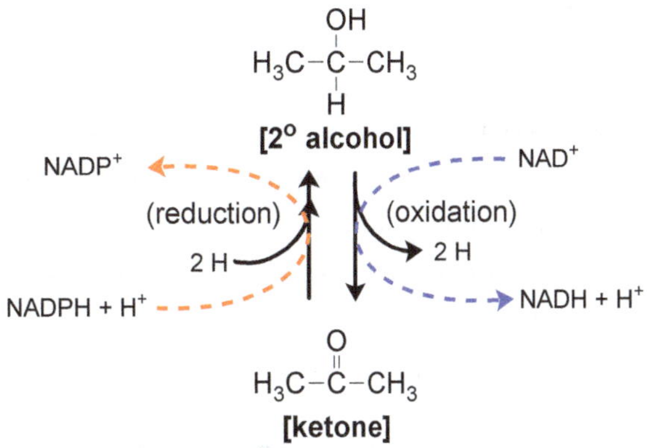

Figure 1.10: The oxidation of a secondary alcohol to a ketone (and the reverse reaction).

The ketone shown in **figure 1.10** is acetone, which is the simplest ketone. Acetone is a key molecule to know (both its name and its molecular structure) because it is the basis of naming of other structures, as well as one of the ketone bodies. Ketones *cannot* be further oxidized. Thus, there are two functionalities to note that cannot be further oxidized: ketones and tertiary (3°) alcohols. However fructose, a ketose sugar, is considered a reducing sugar. A reducing sugar is a sugar that can be oxidized and thus reduces another molecule in the process. Fructose, in an alkaline solution, is in equilibrium with an aldehyde through an enediol intermediate. Thus, fructose still essentially follows the rule that a ketone needs to be isomerized to an aldehyde before becoming oxidized to a carboxylic acid.

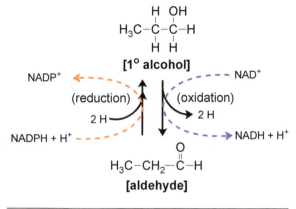

Figure 1.11: The oxidation of a primary alcohol to an aldehyde (and the reverse reaction).

The aldehyde drawn on the left of the oxidation states flowchart (**figure 1.7**) can be produced in two ways. First, consider the reaction of the primary alcohol drawn in **figure 1.11**. Again, compare the structure of the primary alcohol and the aldehyde drawn. Aldehyde functional groups must be on an end carbon (–CH = O, by definition of the molecular structure). Count the number of hydrogens on the primary alcohol and on the aldehyde and note that two hydrogens have been removed, one from the hydroxy group and one from that end carbon itself (again, not implying the actual mechanism). Those two hydrogens must go somewhere—i.e., onto a coenzyme. As mentioned previously, NAD⁺ will be the coenzyme, which gets reduced to NADH + H⁺. To go from a primary alcohol to an aldehyde is an oxidation reaction, indicating that the aldehyde is more oxidized than the alcohol.

The same aldehyde on the flowchart (**figure 1.7**) can be produced by another reaction. In this case start with the ketone drawn (acetone) in the reaction shown in **figure 1.12**. The three-carbon aldehyde and the three-carbon ketone have the same number of carbons, hydrogens, and oxygens. Therefore, these two molecules can be interchanged by an isomerization reaction. There are many enzymes that can simply isomerize a ketone functionality to an aldehyde functionality (and vice versa). For example, fructose (a ketose sugar) and glucose (an aldose, or aldehyde sugar) are both $C_6H_{12}O_6$ molecules. They have the same number of carbons, hydrogens, and oxygens, but they are arranged differently. Thus, fructose and glucose are isomers of one another. It is simply a "switch" of the carbon functionalities on carbons 1 and 2 of these molecules. Thus, aldehydes and ketones, with all component atoms being equal in number, are at the same oxidation level.

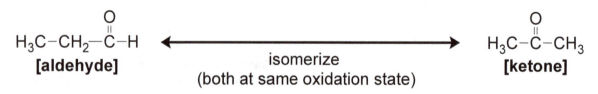

Figure 1.12: The interconversion of aldehydes and ketones.

The last oxidation reaction will get a carbon to the highest oxidation state of carbon—the carboxylic acid group (COOH, or COO⁻). The ultimate oxidation of carbons to carboxylic acids is a *goal* of catabolic pathways. The carboxylic acid group can then be clipped off as carbon dioxide (CO_2) under certain conditions. CO_2 serves as the primary way in which carbon is excreted by breathing it out from the lungs. The carboxylic acid group is the most oxidized state of carbon that is still attached to a molecule. Carbon dioxide is actually the most oxidized state of carbon.

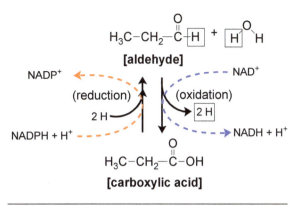

Figure 1.13: The oxidation of an aldehyde to a carboxylic acid (and the reverse reaction).

For the aldehyde to carboxylic acid reaction shown in **figure 1.13**, a molecule of water is required because another oxygen is needed. However, this is *not* a hydration reaction, because only part of the water molecule is added. The structure of the carboxylic acid group, compared to the aldehyde group, looks like an –OH group has replaced the hydrogen of the aldehyde. Thus, count the number of hydrogens on *both* the aldehyde and the water molecule, and compare to the number of hydrogens on the molecule with the carboxylic acid. There are six hydrogens on the molecule with the carboxylic acid. There are eight hydrogens between the molecule with the aldehyde and the water molecule. Thus, there is a loss of two hydrogens—just from two DIFFERENT molecules—one from the aldehyde group and the other from the water molecule. Again, the hydrogens have to go somewhere—to the coenzyme of choice: NAD⁺. NAD⁺ gets reduced to NADH + H⁺. Therefore, the conversion of an aldehyde to a carboxylic acid is an oxidation reaction, coupled with the reduction of a coenzyme.

Before discussing what happens with the carboxylic acid group that has been formed on the molecule, consider now the reverse reactions to go up the oxidation states flowchart. Biochemical pathways can reduce carboxylic acids (and the other functionalities) to return to any of the other oxidation states, including reducing carbons all the way back to an alkane. Typically, this occurs in synthetic pathways. Most synthetic pathways are *reductive* pathways to get carbons to a more reduced state (i.e., fatty acid synthesis, which requires the reduction of keto groups back to alkanes). When going up the oxidation chart, a coenzyme must donate the hydrogens to the molecule to reduce it. The coenzyme of choice: NADPH (reduced form). Recall that NADP⁺/NADPH is the same as NAD⁺/NADH, with the addition of a phosphate "tag." Think of the phosphate tag as a means to indicate that this molecule is to be used for synthetic reactions (going up). NADPH is *not* a donor of hydrogens to the electron transport chain. So everywhere a reduction is shown for the reverse reactions in **figures 1.8**, **1.10**, **1.11**, and **1.13**, NADPH (reduced form) + H⁺ is oxidized to NADP⁺ (as indicated in the reactions going up the page). This *includes* reducing alkenes back to alkanes. A new coenzyme is not needed for this last reduction. The reactions to go from the alcohols to the alkene are dehydration reactions (not redox reactions), which involve the loss of water.

Now that all of the reactions of the oxidation states flowchart have been covered individually, review the entire sequence of reactions in **figure 1.7**. Note, in particular, how the reactions are related to one another going down the oxidation states flowchart (oxidizing), as well as going up the chart (reducing). Also note the relationship of the functional groups to one another, and the coenzymes needed for the reactions. These concepts will be related to the biochemical pathways covered in the remaining chapters of this text to provide the reader with a means to understand the reactions that take place in a particular pathway and why the reactions proceed in a particular order. These are the patterns of biochemical pathways.

There is also a defined protocol for naming the enzymes of the biochemical pathways. While the enzyme naming rules will make more sense when covering the actual enzymes involved in metabolic pathways, a general overview of oxidation-reduction (redox) reaction enzyme naming is presented here. Generally speaking, the easiest way to remember how to name redox enzymes is as follows:

1. Enzymes that use $FAD/FADH_2$ or $NAD^+/NADH$, regardless of which direction the reaction is being carried out (as many redox reactions are reversible), are called *dehydrogenases*. Recall that oxidation is also called dehydrogenation, indicating a loss of hydrogens. Dehydrogenases are *named for the more reduced molecule* because that is the molecule that is being oxidized (dehydrogenated) by the enzyme, as designated by its name.
2. Enzymes that use $NADP^+/NADPH$, regardless of which direction the reaction is being carried out, are called *reductases*. Reductases are *named for the more oxidized molecule* because that is the molecule that is being reduced.

These are the general rules of the oxidation states flowchart and the redox enzyme naming conventions. There will be some exceptions, but if one knows the rules, one just needs to remember the few exceptions (i.e., if "it" was not learned as an exception, "it" must follow the rule).

DECARBOXYLATIONS AND KEY STRATEGIES FOR MAKING ENERGY

Now consider what happens to the carboxylic acid group that is formed from all the oxidation steps previously covered to get from the alkane to the carboxylic acid. Once that carboxylic acid forms, the goal is to clip it off and then transport it through the blood back to the lungs to be breathed off as CO_2. One does not keep every molecule of carbon, oxygen, hydrogen, and so on that is obtained from nutrients in a lifetime. Carbon dioxide is one of the major mechanisms of excreting carbons and oxygens in animals (and other organisms). The problem is that carboxyl acid groups are stable. They are not good leaving groups. A basic rule of organic chemistry—functional groups cannot just be "removed" from a molecule, unless they are good leaving groups. To get rid of the carboxylic acid group as a good leaving group, another reaction will have to be carried out on the molecule to make it a good leaving group.

A. Decarboxylation of carboxylic acid groups (R-COOH → RH + CO_2)

1. Fatty acid (i.e. palmitic acid)

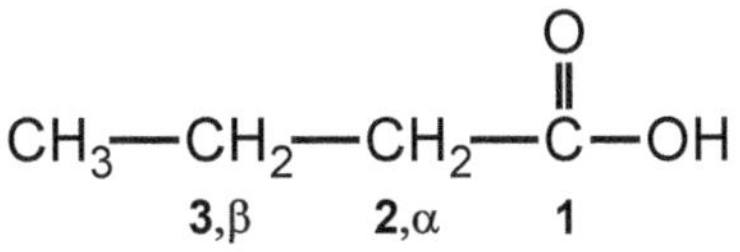

$$CH_3-\underset{3,\beta}{CH_2}-\underset{2,\alpha}{CH_2}-\underset{1}{\overset{O}{\overset{\|}{C}}}-OH$$

is **STABLE**; does NOT decarboxylate

2. α–keto carboxyl

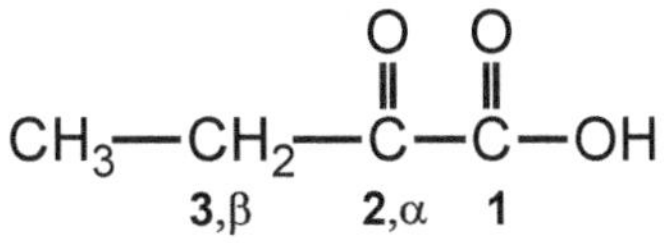

$$CH_3-\underset{3,\beta}{CH_2}-\underset{2,\alpha}{\overset{O}{\overset{\|}{C}}}-\underset{1}{\overset{O}{\overset{\|}{C}}}-OH$$

is **STABLE**; but CAN be decarboxylated with an enzyme

3. β–keto carboxyl

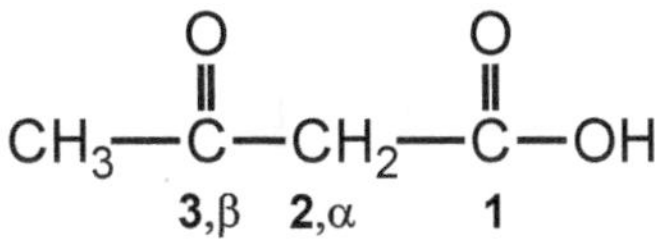

$$CH_3-\underset{3,\beta}{\overset{O}{\overset{\|}{C}}}-\underset{2,\alpha}{CH_2}-\underset{1}{\overset{O}{\overset{\|}{C}}}-OH$$

is **UNSTABLE**; spontaneously decarboxylates

Figure 1.14: Stability of carboxylic acid groups on a fatty acid.

To remove a carboxylic acid group from a molecule (i.e., decarboxylation reaction) in biological pathways, an alpha (α) or beta (β) keto acid is produced. In **figure 1.14** the first molecule drawn is a very short (four-carbon) fatty acid. A fatty acid molecule is very stable, and it does not spontaneously decarboxylate. To make the carboxylic acid group a good leaving group, a keto group is put nearby. The keto group can be put on either the α-carbon (recall that is carbon 2 using scientific numbering) or the β-carbon (carbon 3). If the keto group is placed on the α-carbon (shown in **figure 1.14**), the carboxylic acid group is still stable but can be decarboxylated with an enzyme complex. The basic point is that the decarboxylation of α-keto acids is very "difficult" and requires a large enzyme complex (i.e., the pyruvate dehydrogenase complex and the α-ketoglutarate dehydrogenase complex). While α-keto acids can be decarboxylated, a better option—if possible—is to put the keto group on the β-carbon (shown in **figure 1.14**). Beta-keto acids are unstable and will spontaneously decarboxylate. Note that "spontaneous" does not mean "instantaneous." Spontaneous does not indicate anything about the reaction rate. The decarboxylation of β-keto acids still generally requires an enzyme to move the reaction along in most biological pathways. For β-keto acids the carboxylic acid is now a good leaving group and can "leave" the molecule without the need of an enzyme. One example is the ketone body acetoacetate (a β-keto acid), which can spontaneously decarboxylate to form acetone, another ketone body. Ketone bodies are an important water-soluble fuel source for the body derived from the catabolism of fatty acids. Acetone, though, can only be excreted. It cannot be used as a fuel source, as will be discussed in chapter 6.

B. Key strategies for making energy (ATP)

GOAL: Make NADH and $FADH_2$ for ATP production by the electron transport chain. The following overall scheme for making energy is a *very* simplistic view and is based on patterns—not mechanisms. The goal of catabolic processes is to oxidize the carbons of molecules by removing hydrogens (dehydrogenation). Remember FAD as the coenzyme for the oxidation of alkanes to alkenes, because FAD can accept *two* hydrogens from the *two* different carbons of the alkane to be reduced to $FADH_2$. (Note, the actual reason is that the reducing power of alkanes is not sufficient to reduce NAD^+ to NADH). For the oxidation of alcohols and aldehydes, the coenzyme used is NAD^+ because the reactions remove hydrogen(s) from *one* carbon, and NAD^+ can only accept *one* hydrogen atom (as a hydride ion carrying both electrons). *Also note*: The oxidation of aldehydes generally yields a high-energy bond "S—P," which can be used for *substrate-level phosphorylation* to yield ATP (or an energy equivalent).

As noted, the carboxylic acid is the most oxidized state of carbon on a molecule. Notice on the oxidation chart that every time a molecule was oxidized, a coenzyme was reduced. Again, oxidation (going down the page) is what catabolic pathways accomplish. The more oxidations needed to get a carbon to the carboxylic acid functionality, the more $FADH_2$ and NADH produced. $FADH_2$ and NADH are the substrates for the electron transport chain, which is the major producer of ATP for a cell. The more $FADH_2$ and NADH produced by a catabolic pathway, the more ATP that can be made by the electron transport chain. This is the goal of catabolism: to make reducing power ($FADH_2$ and NADH) produced by oxidation reactions that in turn can be used by the electron transport chain to make ATP. Also, the carboxylic acids ultimately produced by these catabolic processes deprotonate. These protons must be buffered by components in the blood to prevent significant changes in the pH of blood. In addition, the products of the catabolic pathways are waste products that can be excreted (i.e., CO_2, H_2O, and NH_4^+ for nitrogen-containing molecules).

The more reduced the molecule, the more oxidations that have to be done to get each carbon up to a carboxylic acid. Thus, the more reduced molecules ultimately provide more $FADH_2$ and NADH to the electron transport chain for ATP production. Hence, fatty acids will ultimately produce more ATP per molecule than a molecule of glucose (sugar). The carbons on sugars like glucose (six carbons) are alcohols or aldehydes and do not require as many oxidations to get rid of the six carbons of a molecule of glucose. For a fatty acid like palmitate (a C_{16} fatty acid), there are fifteen carbons at the level of an alkane, and one carbon is a carboxylic acid as shown in **figure 1.1**. Therefore, many more oxidations are required to get rid of all the carbons of a molecule of palmitate. Hopefully, after covering carbohydrate versus fatty acid catabolism in subsequent chapters of this book, the reader will truly understand why a fatty acid molecule will ultimately yield more ATP than a molecule of glucose.

NAMING OF ENZYMES

The general classifications and rules for naming enzymes are outlined below. These rules will be covered in more detail using the actual enzyme names of reactions of the metabolic pathways covered in this text.

I. Six Functional Classification Groups of Enzymes
   A. Oxidoreductases
      1. Catalyze oxidation-reduction reactions (a.k.a. redox reactions).
      2. *Oxidation*—**removal of electrons (molecule becomes more +).**
         Reduction—**addition of electrons (molecule becomes more −) Note:** For readers who have difficulty seeing an "addition" of something as a "reduction," electrons are **negatively** charged, so the "addition" of electrons to a molecule makes the charge **more negative** (or less positive)—i.e., "**reduced.**"
      3. These enzymes are called **dehydrogenases** = removal of hydrogens or **reductases** = addition of hydrogens.
         (Hydrogen is the carrier of the electrons, usually; an exception—the electron transport chain complexes.)
      4. The enzymes require a **coenzyme**, which acts as a donor or acceptor of the electrons (carried by hydrogens; again *except* electron transport chain complexes—many of the complexes use various metal ions as coenzymes for electron movement).
   B. Transferases
      1. Transfer a functional group from one compound to another compound.
      2. Example: a **kinase** = an enzyme that transfers phosphate (PO_4) groups.
   C. Hydrolases
      1. Break bonds by the addition of H_2O.
      2. Dehydratases—remove H_2O from a compound(s).
   D. Lyases
      1. Break bonds without the addition of H_2O.
      2. Forms double bonds (*note*—double bonds also can be formed by oxidation-reduction reactions).
   E. Isomerases
      1. Catalyze the formation of isomers of a compound (configurational change).
      2. **Isomers** = same number of atoms and bonds, but in a different configuration.
   F. Ligases
      1. Make bonds using energy (ATP).
   G. Note
      1. Transferases and ligases are two classes of enzymes that can make *big* molecules (i.e., make bonds).
      2. Hydrolases and lyases are two classes of enzymes that break bonds.

II. Rules for Naming Enzymes
   A. Suffix for enzymes is **−ase**

B. Anabolic (Synthetic) Reactions
   1. Name **product**.
   2. Followed by either (a) **Synthase**—no energy required (classified—*transferase*) OR (b) **Synthetase**—energy required (classified—*ligase*).
C. Catabolic (Breakdown) Reactions
   1. Name reactant(s).
   2. Followed by the type of reaction being carried out.
D. Oxidoreductase Enzymes (a.k.a. Redox Enzymes)
   1. For ***CATABOLIC*** pathways:
      a. Name the **reduced** molecule
      b. Followed by *dehydrogenase*
      c. **UNLESS** molecular oxygen is used directly, in which case you follow it by *oxidase*.
      d. **Note:** When multiple reactions are carried out by an enzyme, if any one of the reactions is redox, then the enzyme will be called a dehydrogenase.
      e. **Tip:** MOST (but not all) dehydrogenases use NAD^+/NADH or FAD/$FADH_2$, regardless of the direction of the actual reaction (as many oxidation/reduction reactions are REVERSIBLE).
   2. For ***SYNTHETIC*** pathways:
      a. Name the **oxidized** molecule
      b. Followed by *reductase*
      c. **Tip:** MOST (but not all) reductases use $NADP^+$/NADPH, regardless of the direction of the actual reaction.
   3. Other exceptions: The electron transport chain complexes have several names, some of which also use the name *reductase*.
E. Other Rules for Naming Enzymes
   1. **Phosphatases** (classified—hydrolase)—hydrolyze off phosphate groups.
   2. **Phosphorylases**—put phosphate groups on (classified—transferase).
   3. **Mutases**—isomerases that catalyze the **intra**molecular rearrangement of functional groups (i.e., in glycolysis, *phosphoglycerate mutase* moves a phosphate group from carbon 3 to carbon 2 of the molecule).
   4. **Kinases** (classified—transferase)—transfer phosphate groups to/from ATP. Kinases are always named for the **molecule** that would be **accepting the phosphate group** from ATP. In other words, as if ATP is always the DONOR of the phosphate group.
      a. Thus, if ATP is actually donating a phosphate group in a reaction (i.e., one of the substrates of the reaction):
         (1) Name the **SUBSTRATE** (the acceptor of the phosphate group)
         (2) Followed by *kinase*
      b. If ATP is a product of a reaction:
         (1) Name the **PRODUCT** (the one that would be accepting the phosphate group from ATP if the reaction ran in the reverse direction)
         (2) Followed by *kinase*

c. For example: in glycolysis there are four kinases:
   (1) In the first half of glycolysis (where ATP is used), the first two kinases are named for the **substrates** of the reactions: <u>hexo</u>kinase and <u>phosphofructo</u>kinase.
   (2) In the second half of glycolysis (where ATP is produced), the kinases are named for the **products** of the reactions: <u>phosphoglycerate</u> kinase and <u>pyruvate</u> kinase.

5. **Aminotransferase** (formerly called transaminase; classified transferase)—transfers amino groups.
   a. **New nomenclature names the aminotransferase after the *donor* amino acid.**
   b. Old nomenclature names the product amino acid first, then the product α-keto acid, followed by transaminase.
   c. Examples:
      (1) Rxn: alanine + α-ketoglutarate ↔ pyruvate + glutamate
         alanine aminotransferase (ALT)—new nomenclature
         glutamate/pyruvate transaminase (GPT)—old nomenclature
      (2) Rxn: aspartate + α-ketoglutarate ↔ oxaloacetate + glutamate
         aspartate aminotransferase (AST)—new nomenclature
         glutam**a**te/oxaloacetate transaminase (GOT)—old nomenclature
   d. Clinical relevance: test for serum levels of these transaminases liver damage → elevated levels of ALT and AST (example: old nomenclature may see as SGOT for "serum" GOT)

6. **Hydrolase** enzymes:
   a. Name reactant
   b. Plus attach **–ase** ending (example: sucr<u>ase</u>—hydrolyzes sucrose)
      Note: Digestive enzymes are *hydrolases*.

Note
There are often exceptions to these rules, but it is easier to note the exceptions when they occur if one knows the general rules for naming enzymes.

The Nomenclature Committee of the International Union of Biochemistry and Molecular Biology (NC-IUBMB) and the IUPAC-IUBMB Joint Commission on Biochemical Nomenclature continually make recommendations and update the rules for naming enzymes (http://www.chem.qmul.ac.uk/iubmb/enzyme/rules.html; http://www.chem.qmul.ac.uk/iubmb/enzyme/). These include systematic naming conventions, which often result in longer and more cumbersome names for enzymes. Updates to the common or "trivial" naming rules of enzymes have also been introduced. However, for enzymes of metabolic pathways that have been known for some time (i.e., the pathways covered in this text), the enzyme names assigned, based on the naming rules in place at the time of their discovery, are how the enzymes are best known. The rules outlined above still serve as the primary basis for how the metabolic enzyme names can be deduced from the reactions they catalyze.

PROBLEM SET: BASIC PRINCIPLES OF OXIDATION STATES

IMPORTANT: Please do these problems and exercises by using your notes from this oxidation states chapter, which will provide more insight and comprehension than simply looking at the answers.

Exercise 1

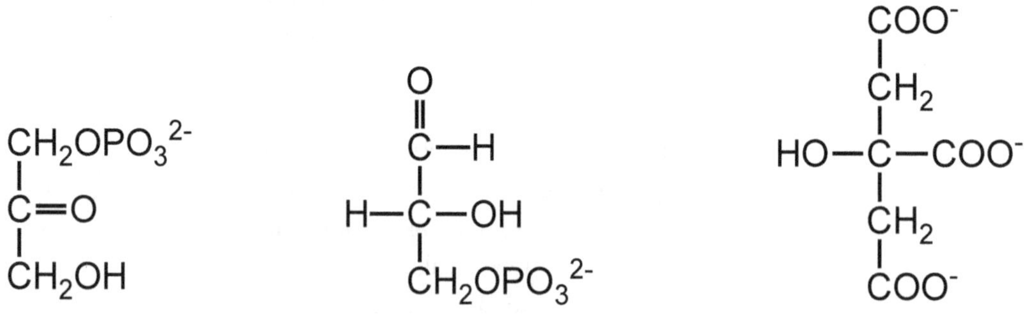

molecule 1

$CH_2OPO_3^{2-}$
|
$C=O$
|
CH_2OH

molecule 2

O
‖
$C-H$
|
$H-C-OH$
|
$CH_2OPO_3^{2-}$

molecule 3

COO^-
|
CH_2
|
$HO-C-COO^-$
|
CH_2
|
COO^-

Figure 1.15: Comparison of functional groups.

1. Which of the three molecules in **figure 1.15** have functional groups that CANNOT be further oxidized? (Note, they all contain multiple functional groups, but which functional groups cannot be further oxidized?)
    A. molecule 1
    B. molecule 2
    C. molecule 3
    D. molecules 1 and 3
    E. molecules 1, 2, and 3

Exercises 2 and 3

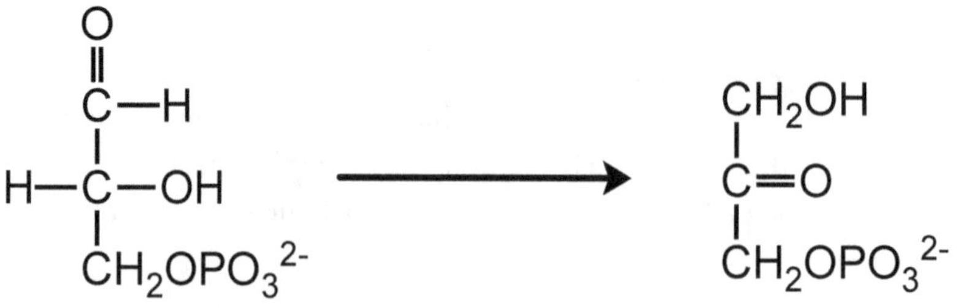

1

O
‖
$C-H$
|
$H-C-OH$
|
$CH_2OPO_3^{2-}$

→

2

CH_2OH
|
$C=O$
|
$CH_2OPO_3^{2-}$

Figure 1.16: Identification of the type of reaction and its enzyme.

2. In the reaction drawn in **figure 1.16**, what has happened to molecule 1 to convert it to molecule 2?
   A. phosphorylation
   B. dehydration
   C. isomerization
   D. oxidation
   E. reduction

3. For the reaction drawn in **figure 1.16**, what type of enzyme would carry out this reaction?
   A. epimerase
   B. isomerase
   C. dehydrogenase
   D. kinase
   E. hydrolase

Exercise 4

1. On a separate sheet of paper, redraw the four molecules in **figure 1.17** in the order that a **metabolic pathway** would need to follow to go *from* **the most reduced molecule** (draw at the top of the page) *to* **the most oxidized molecule** (draw toward the bottom of the page). Leave space between the molecules to do the following steps. (**Hint:** Focus on the differences between the four molecules; if the only difference between molecules is the number of hydrogens, then the molecule with *more hydrogens* is the *more reduced* molecule.)

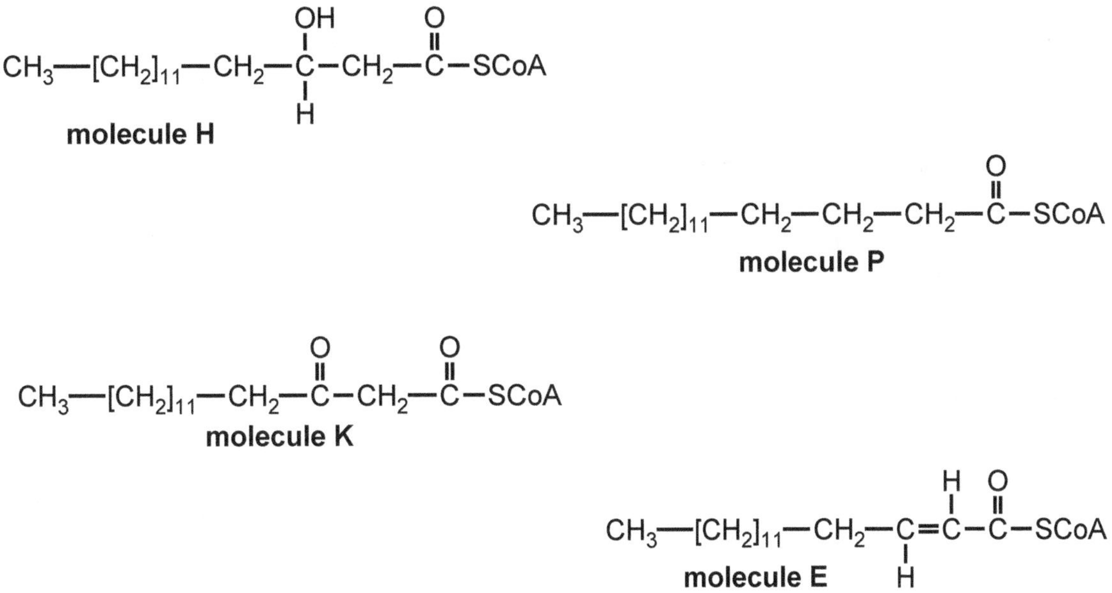

CH_3—$[CH_2]_{11}$—CH_2—$\underset{H}{\overset{OH}{C}}$—$CH_2$—$\overset{O}{\underset{}{C}}$—SCoA

molecule H

CH_3—$[CH_2]_{11}$—CH_2—CH_2—CH_2—$\overset{O}{\underset{}{C}}$—SCoA

molecule P

CH_3—$[CH_2]_{11}$—CH_2—$\overset{O}{\underset{}{C}}$—$CH_2$—$\overset{O}{\underset{}{C}}$—SCoA

molecule K

CH_3—$[CH_2]_{11}$—CH_2—$\overset{H}{\underset{}{C}}$=$\underset{H}{\overset{}{C}}$—$\overset{O}{\underset{}{C}}$—SCoA

molecule E

Figure 1.17: Exercise 4. Reduced to oxidized reaction sequence. Note, the four molecules have been "named" using random letters.

2. Now draw reaction arrows between the molecules. (Since there are 4 molecules, you should draw three reaction arrows, indicating the three reactions that must take place to go from the most reduced molecule to the most oxidized molecule.) Identify these three types of reactions (i.e., oxidation/reduction and so on).
3. Now indicate the side reactants/products that are necessary for each of the three reactions.
4. Using the appropriate assigned molecule letter, name the enzyme that would carry out each of the three reactions. You need to know whether you are naming the enzyme for the reactant or product of the given reaction. Review the "Naming of Enzymes" section of the chapter, which provides some rules for how to name the enzymes. (For example: "molecule z dehydrogenase," as an enzyme that carries out an oxidation-reduction reaction. In this case molecule z must be the *reduced* molecule in the reaction because that is how dehydrogenases are named.)

Exercise 5

1. On a separate sheet of paper, redraw the four molecules in **figure 1.18** in the order that a **metabolic pathway** would need to follow to go ***from* the most oxidized molecule** (draw at the top of the page) ***to* the most reduced molecule** (draw toward the bottom of the page). Leave space between the molecules to do the following steps. (**Hint:** Focus on the differences between the four molecules; if the only difference between molecules is the number of hydrogens, then the molecule with *more hydrogens* is the *more reduced* molecule.)
2. Now draw reaction arrows between the molecules. (Since there are 4 molecules, you should draw three reaction arrows, indicating the three reactions that must take place to go from the most oxidized molecule to the most reduced molecule.) Identify these three types of reactions (i.e., oxidation/reduction and so on).

$CH_3-\overset{H}{\underset{H}{C}}=\overset{}{\underset{}{C}}-\overset{O}{\underset{}{C}}-S-ACP$ **molecule N**

$CH_3-\overset{O}{\underset{}{C}}-CH_2-\overset{O}{\underset{}{C}}-S-ACP$ **molecule T**

$CH_3-CH_2-CH_2-\overset{O}{\underset{}{C}}-S-ACP$ **molecule A**

$CH_3-\overset{OH}{\underset{H}{C}}-CH_2-\overset{O}{\underset{}{C}}-S-ACP$ **molecule Y**

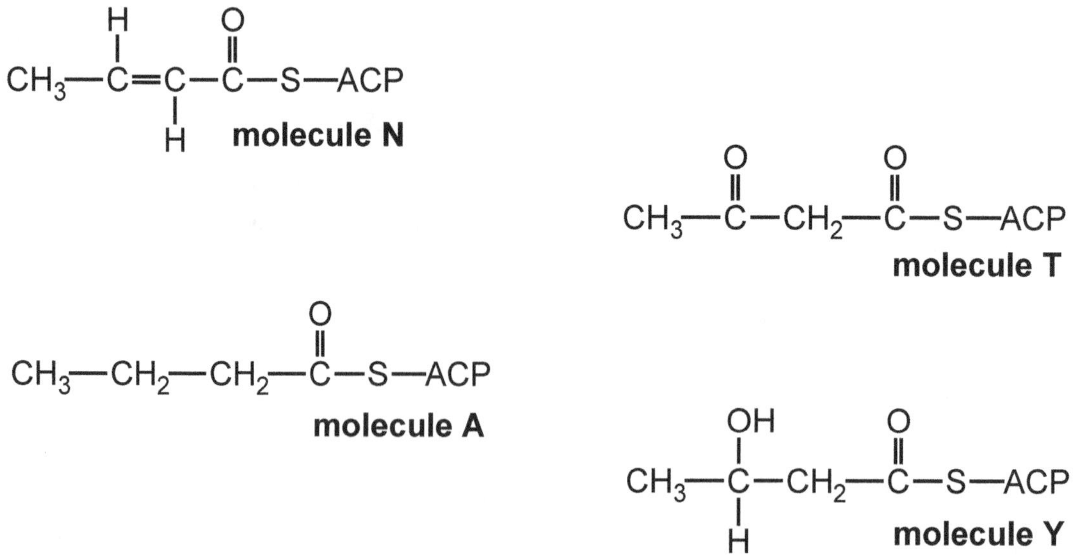

Figure 1.18: Exercise 5. Oxidized to reduced reaction sequence. Note, the four molecules have been "named" using random letters.

3. Now indicate the side reactants/products that are necessary for each of the three reactions.
4. Using the appropriate assigned molecule letter, name the enzyme that would carry out each of the three reactions. You need to know whether you are naming the enzyme for the reactant or product of the given reaction. Review the "Naming of Enzymes" section, which provides some rules for how to name the enzymes. (For example: "molecule z reductase," as an enzyme that carries out an oxidation-reduction reaction. In this case molecule z must be the *oxidized* molecule in the reaction because that is how reductases are named.)

Exercise 6

1. Draw in the functional groups on the top carbons of molecules B, C, and D in **figure 1.19** that would be required to convert **molecule A** to **molecule E**. (**Hint:** Again, focus on the differences between molecule A and molecule E.)

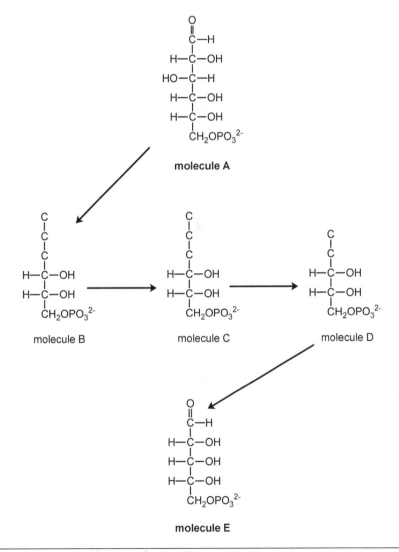

Figure 1.19: Exercise 6. Determination of a reaction sequence.

PROBLEM SET: SOLUTIONS

1. (D) Molecule 1 has a ketone functionality, and molecule 3 has a 3° alcohol functionality, neither of which can be further oxidized.
2. (C) Isomerization of an aldehyde to a ketone
3. (B) Isomerase

Exercise 4 solution
(See **figure 1.20**.)

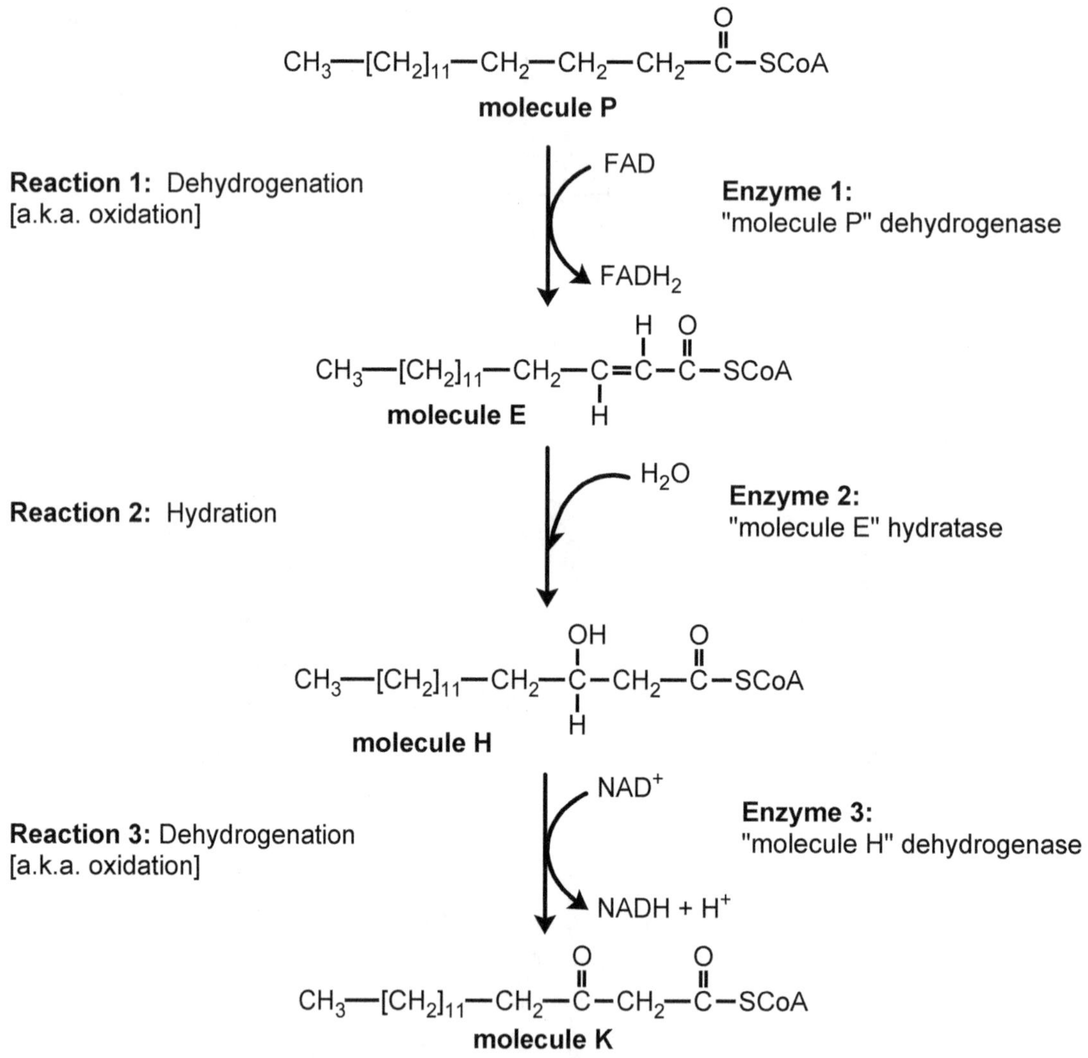

$$CH_3-[CH_2]_{11}-CH_2-CH_2-CH_2-\overset{O}{\underset{\|}{C}}-SCoA$$
molecule P

Reaction 1: Dehydrogenation [a.k.a. oxidation]

↓ FAD → FADH$_2$

Enzyme 1: "molecule P" dehydrogenase

$$CH_3-[CH_2]_{11}-CH_2-\underset{H}{\overset{H}{\underset{|}{C}}}=\overset{}{\underset{H}{C}}-\overset{O}{\underset{\|}{C}}-SCoA$$
molecule E

Reaction 2: Hydration

↓ H$_2$O

Enzyme 2: "molecule E" hydratase

$$CH_3-[CH_2]_{11}-CH_2-\underset{H}{\overset{OH}{\underset{|}{C}}}-CH_2-\overset{O}{\underset{\|}{C}}-SCoA$$
molecule H

Reaction 3: Dehydrogenation [a.k.a. oxidation]

↓ NAD$^+$ → NADH + H$^+$

Enzyme 3: "molecule H" dehydrogenase

$$CH_3-[CH_2]_{11}-CH_2-\overset{O}{\underset{\|}{C}}-CH_2-\overset{O}{\underset{\|}{C}}-SCoA$$
molecule K

Figure 1.20: Exercise 4 solution.

For reactions 1 and 3: Oxidation and reduction reactions are ALWAYS coupled. However, by labeling reactions 1 and 3 as oxidation reactions, the reactions refer to what is happening in going from molecule P to molecule E (reaction 1) and molecule H to molecule K (reaction 3). Again, focus on what is DIFFERENT about each of these molecules. Many molecules have more than one functional group—one needs to recognize what is changing about the molecules in the reactions to figure out what type of reaction is happening and the type of enzyme that will carry out the reaction.

In **reaction 1**, carbons 2 and 3 of molecule P are at the level of an alkane. In molecule E there is a double bond between carbons 2 and 3. Therefore, the molecule has been oxidized to form molecule E. Thus, a coenzyme must be reduced. FAD is typically used when oxidizing a molecule from the level of an alkane to an alkene. So FAD is reduced to $FADH_2$. Enzymes that carry out oxidation-reduction reactions in generally "catabolic" pathways (i.e., going "down" the oxidation chart) are *dehydrogenases*—which are named for the more REDUCED molecule. Molecule P (alkane level) is the more reduced molecule of the two, so the enzyme name is "molecule P" dehydrogenase.

Reaction 2 is a hydration of the alkene of molecule E to form molecule H. This is not an oxidation reaction. It is simply a hydration, the addition of H_2O across the double bond—hence, H_2O is a second reactant necessary for this reaction. This enzyme name would not be obvious because there are no hard and fast rules for naming enzymes that carry out hydration and dehydration reactions. However, these enzymes are typically named in some fashion after the molecule that is being hydrated (*hydratase*) or dehydrated (*dehydratase*)—though these are typically reversible reactions. In this case the reaction is hydrating molecule E. So the enzyme is called "molecule E" hydratase.

In **reaction 3**, carbon 3 of molecule H has an alcohol group (-OH group) on it. Molecule K has a keto group on carbon 3. Thus, this reaction is also an oxidation reaction (a.k.a. dehydrogenation). NAD^+ is typically used for the oxidations of secondary alcohols to ketones (and for oxidations of primary alcohols to aldehydes, as well as aldehydes oxidized to carboxylic acids). Thus, NAD^+ is reduced to $NADH + H^+$. Again, the enzyme would be a *dehydrogenase* named for the more REDUCED molecule. The alcohol group of molecule H is more reduced than the keto group on molecule K (again, focus on the differences between molecules). Thus, the enzyme would be "molecule H" dehydrogenase.

Exercise 5 solution
(See **figure 1.21**.)

For reactions 1 and 3: Oxidation and reduction reactions are ALWAYS coupled. However, by labeling reactions 1 and 3 as reduction reactions, the reactions refer to what is happening in going from molecule T to molecule Y (reaction 1) and molecule N to molecule A (reaction 3). Again, focus on what is DIFFERENT about each of these molecules. Many molecules have more than one functional group. One needs to recognize what is changing about the molecules in the reactions to figure out what type of reaction is happening and the type of enzyme that will carry out the reaction.

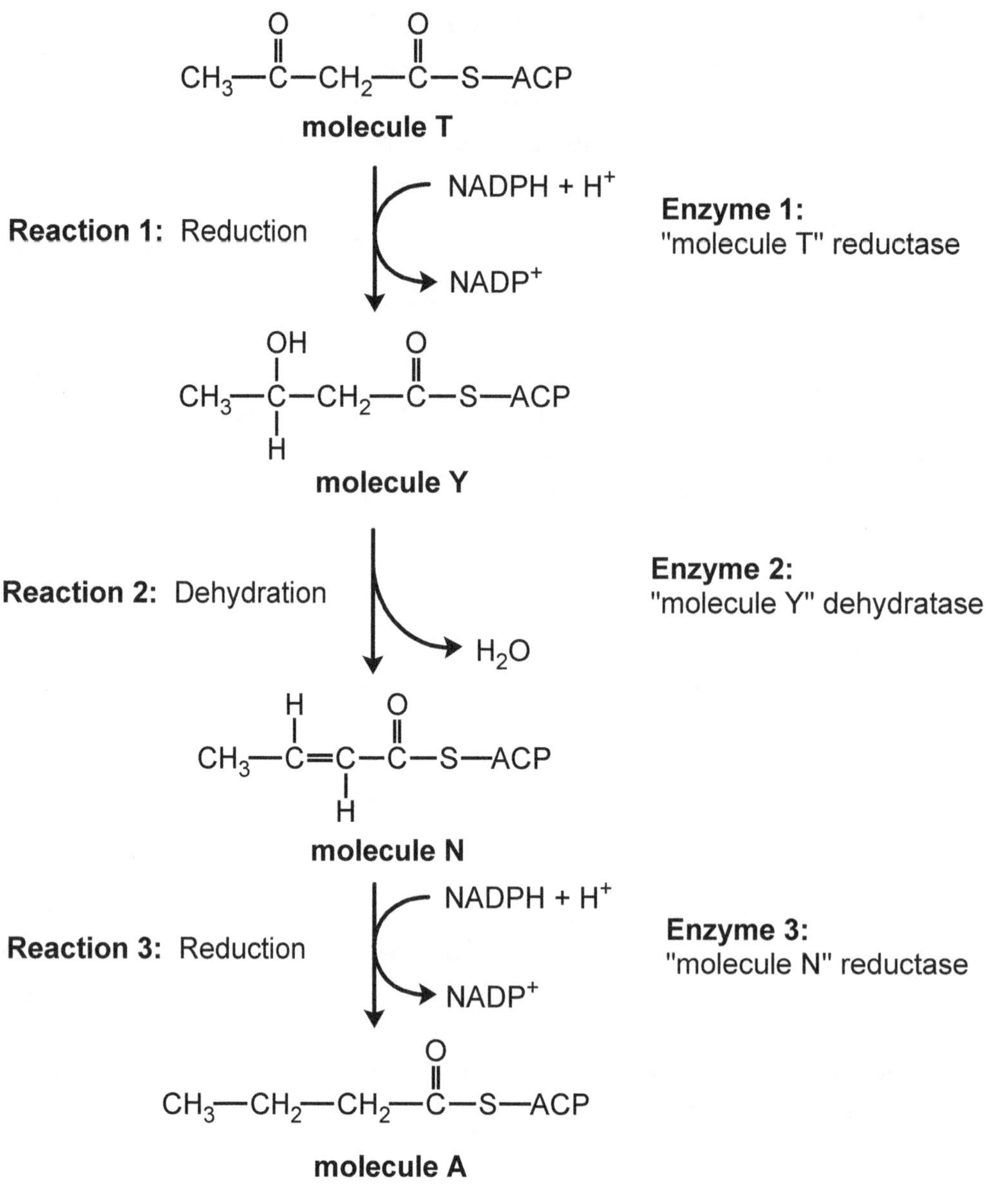

$$CH_3-\overset{\overset{O}{\|}}{C}-CH_2-\overset{\overset{O}{\|}}{C}-S-ACP$$
molecule T

Reaction 1: Reduction — NADPH + H$^+$ → NADP$^+$

Enzyme 1: "molecule T" reductase

$$CH_3-\underset{\underset{H}{|}}{\overset{\overset{OH}{|}}{C}}-CH_2-\overset{\overset{O}{\|}}{C}-S-ACP$$
molecule Y

Reaction 2: Dehydration → H$_2$O

Enzyme 2: "molecule Y" dehydratase

$$CH_3-\underset{\underset{H}{|}}{\overset{\overset{H}{|}}{C}}=C-\overset{\overset{O}{\|}}{C}-S-ACP$$
molecule N

Reaction 3: Reduction — NADPH + H$^+$ → NADP$^+$

Enzyme 3: "molecule N" reductase

$$CH_3-CH_2-CH_2-\overset{\overset{O}{\|}}{C}-S-ACP$$
molecule A

Figure 1.21: Exercise 5 solution.

In **reaction 1**, molecule T has a keto group on carbon 3, while carbon 3 of molecule Y has an alcohol group (–OH group or hydroxy group) on it. Thus, this reaction is a reduction reaction, and a coenzyme must also be oxidized. NADPH + H$^+$ is typically used for *all* reduction reactions in synthetic pathways. Thus, NADPH + H$^+$ is oxidized to NADP$^+$. The enzyme would be a *reductase* named for the more OXIDIZED molecule. The keto group of molecule T is more oxidized than the hydroxy group on molecule Y (again, focus on the differences between molecules). Thus, the enzyme would be "molecule T" reductase.

Reaction 2 is a dehydration reaction removing the hydroxy group on carbon 3 and a hydrogen on carbon 2 of molecule Y to form the alkene of molecule N. This is not an oxidation-reduction reaction. It is simply a dehydration reaction, the removal of H_2O to form the double bond (recall alcohols and alkenes are at the same oxidation level). This enzyme name would not be obvious because there are no hard and fast rules for naming enzymes that carry out hydration and dehydration reactions. However, these enzymes are typically named in some fashion after the molecule that is being hydrated (*hydratase*) or dehydrated (*dehydratase*)—though these are typically reversible reactions. In this case the reaction is dehydrating molecule Y. So the enzyme is called "molecule Y" dehydratase.

In **reaction 3**, molecule N has a double bond between carbons 2 and 3, while both carbons 2 and 3 of molecule A are at the level of an alkane. Therefore, molecule N has been reduced to form molecule A, and a coenzyme must also be oxidized. $NADPH + H^+$ is typically used for *all* reduction reactions in synthetic pathways. Thus, $NADPH + H^+$ is oxidized to $NADP^+$. The enzyme would be a *reductase* named for the more OXIDIZED molecule. The alkene group of molecule N is more oxidized than the alkane carbons on molecule A (again, focus on the differences between molecules). Thus, the enzyme would be "molecule N" reductase.

Exercise 6 solution
(See **figure 1.22**.)
There are several possibilities for the solution to this problem (one possibility is illustrated in **figure 1.22**). The primary differences one needs to notice between molecules A and E is that carbon 1 of molecule A is removed and that carbon 2 of molecule E (which was carbon 3 of molecule A) has the –OH group pointing in the opposite direction. The orientation of functional groups around chiral carbons is significant.

The two main reactions one needs to include are the oxidation of the aldehyde carbon 1 of molecule A to a carboxylic acid group and the oxidation of an –OH group on either carbon 2 or carbon 3 of molecule A to a keto group (creating either an α-keto acid or a β-keto acid). These two reactions are necessary for the loss of carbon 1 from molecule A.

While generally in metabolism the oxidation to the acid would occur first, for this exercise it does not matter which oxidation one put first—just as long as both oxidations were included.

If one chose to make an α-keto acid, molecule D would have been an aldose after the decarboxylation. However, carbon 2 would have the –OH in the wrong configuration—thus, an epimerase would be needed to convert molecule D to molecule E.

If one chose to make a β-keto acid as shown in **figure 1.22** (preferred metabolically because β-keto acids are easier to decarboxylate), molecule D would be a ketose after the decarboxylation, and an isomerase would be needed to convert molecule D to molecule E.

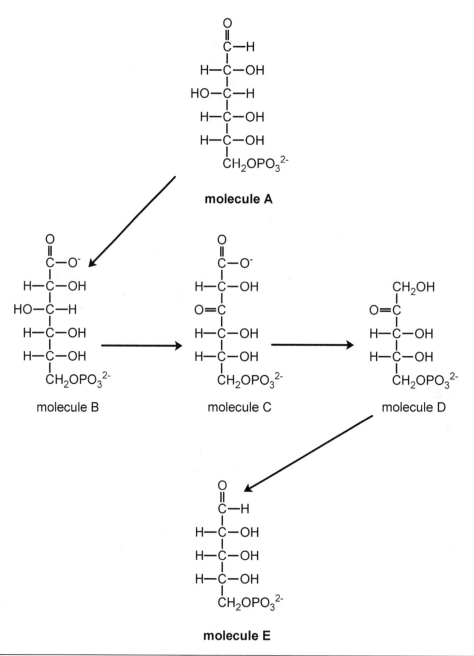

Figure 1.22: Exercise 6 solution.

CHAPTER 2
METABOLISM OVERVIEW AND GLYCOLYSIS

OBJECTIVES

1. Define the process of glycolysis.
2. Explain the purpose of the pathway of glycolysis.
   a. Account for the net yield of ATP in the catabolism of glucose to pyruvate; which reactions use and which reactions produce ATP.
3. Identify where glycolysis takes place in the cell.
4. Explain how glycolysis is carried out in the cell.
   a. Name and recognize the structures of the starting materials, key intermediates, and end products.
   b. Name the enzymes involved in the aerobic and anaerobic catabolism of glucose and identify the reactions they catalyze.
   c. Identify the types of glycolytic reactions that are carried out by kinases and dehydrogenases (and how they are named), isomerases, mutases, aldolase, and enolase.
   d. Describe the purpose of the glucose to glucose 6-phosphate reaction and identify the two isozymes that carry out this reaction.
   e. Describe the purpose of the last three steps (steps 8, 9, and 10) of glycolysis in respect to producing the end products ATP and pyruvate.
   f. Define the term "substrate level phosphorylation" and identify the intermediates and reactions of glycolysis that fit this definition.
5. Explain when glycolysis takes place.
   a. Identify the major regulatory steps in glycolysis, the enzymes that catalyze these steps, and the molecules that regulate them.
   b. Identify the key reaction that differentiates aerobic from anaerobic glycolysis and why oxygen availability controls whether or not this additional reaction is carried out.
6. Identify the fates of each of the products of glycolysis, under both aerobic and anaerobic conditions.
   a. Describe the difference in how cytosolic NAD^+ is regenerated for glyceraldehyde 3-phosphate dehydrogenase during aerobic versus anaerobic glycolysis.
   b. Identify the link between glyceraldehyde 3-phosphate dehydrogenase and lactate dehydrogenase during anaerobic glycolysis.

    c. Describe why shuttles are necessary to carry the electrons via hydrogens (reducing equivalents) between the cytoplasm and the mitochondria under aerobic conditions and list the two different types of shuttles.
7. Describe the physiological role for the different affinities for glucose (K_m values), rates (V_{max} values), and regulation for glucokinase versus hexokinase.

OVERVIEW OF METABOLISM

There are four major classes of biological molecules (see **table 2.1**): carbohydrates (or sugars), lipids, proteins, and nucleic acids (DNA and RNA). These large "functioning biomolecules" are polymers or are polymer-like and are made from monomers termed "basic building blocks." Glycogen is a polymer of glucose monomers. Proteins are polymers made up of amino acid monomers. DNA and RNA are polymers made up of nucleotide monomers. Triacylglycerols (a.k.a. triacylglycerides, triglycerides, or fat) are polymer-like. A triacylglycerol has three fatty acid monomers attached to a glycerol backbone.

Table 2.1: Four classes of biomolecules

FUNCTIONING BIOMOLECULE	ACTIVATED PRECURSOR	BASIC BUILDING BLOCK
1. Carbohydrate [i.e. glycogen]	UDP-glucose	Monosaccharide
2. Lipids (Fat) [i.e. triacylglycerols and cholesterol]	Fatty acyl-CoA	Fatty acids
3. Proteins	tRNA-amino acid	Amino acids
4. Nucleic acids [i.e. DNA and RNA]	dNTPs and NTPs	Ribo- (and deoxyribo-) nucleic acids

The ability to attach the basic building blocks together to make these larger functioning biomolecules requires energy and usually reducing power in the form of NADPH. Thus, the synthesis of these functioning biomolecules generally goes through an activated precursor stage, which harnesses the energy needed to create bonds between the basic building blocks. For instance, monosaccharides are typically attached to uridine triphosphate (UTP) to form a UDP-sugar (i.e., UDP-glucose). The removal of the UDP unit provides the energy for the attachment of the sugar to the larger molecule (as in the synthesis of glycogen or a glycolipid).

The metabolic pathways described in this text will focus on catabolism (breakdown) and anabolism (synthesis) of the monomers or basic building blocks themselves. In particular, the focus will be on the pathways associated with the catabolism and synthesis of glucose and fatty acids. The nutrients one obtains in the diet are broken down to these basic building blocks and ultimately absorbed by various tissues. These basic building blocks can be catabolized further by cells or used for other synthetic processes. The body also has the ability to make some of these basic building blocks de novo (i.e., "from scratch") in cells.

HOW TO STUDY METABOLIC PATHWAYS

When one is describing an event to someone else, there are generally six key questions one should answer to relay the complete story of the event: Who? What? Why? Where? When? How? Answering these six questions is also essential to developing an understanding of metabolic pathways.

The question *who* refers to what organism is being studied. For this text the answer to *who* is humans, but many of the pathways covered are found in other animals, plants, and microorganisms. Bacteria inhabit nearly every environmental niche on earth. As these bacteria are exposed to temperatures, pH, and nutrients unique to their environments, bacteria can catalyze many different types of reactions that humans cannot do within their normal temperature and pH ranges. However, bacteria still have to carry out many of the basic types of reaction sequences needed for the catabolism and anabolism of their functioning biomolecules.

All pathways in this text will provide an answer to four of the basic questions: *what, why, where,* and *how*. The question *what* refers to the definition of the pathway. The question *why* is the purpose or function of the pathway. *Where* describes the location of the pathway in a cell and/or tissue(s). Some pathways only take place, for instance, in the liver. Ketone body synthesis is an example of a pathway that only takes place in the liver. *How* is a question that seems to encompass everything else regarding a pathway, including the molecules, the reactions, the enzymes and their mechanisms. There are key points, though, that one should know about how a pathway occurs that will help keep a big picture perspective on what the pathway needs to accomplish. Learning key structures of pathways will help one form a picture in one's mind of the overall sequence of the pathway. By learning how pathways follow the basic recipe of metabolism, then one can simply learn the "list" of the reaction sequences. If one truly understands how to apply these general patterns and rules, one can draw out or recognize any step based on a knowledge of key structures, oxidation state sequences, and enzyme naming rules.

In beginning to understand how the pathway occurs, start with knowing the starting and ending products of the pathway, including their structures. This is the picture of what the pathway is starting with and trying to get to. Know any steps that make or require ATP, as these are often important steps of a pathway. Know the vitamins and coenzymes necessary for particular enzymes. Coenzymes are typically derived from vitamins. Patients suffering from particular vitamin deficiencies display clinical signs and symptoms based on the enzymes affected.

The question *when* will be answered as one begins to understand how various metabolic pathways fit together. The answer to the question *when* demonstrates one's understanding of the integration of metabolic pathways. Answering the question *when* for a particular pathway includes indicating under what conditions the pathway should be on (active) or off (inhibited); the enzymes that are regulated for the pathway and what regulates them—their activators and/or inhibitors; what other pathways need to run simultaneously or in sequence; and what pathways will never be on in the *same cell* at the same time. Answers to the question *when* will be provided in the context of the pathways covered. However, answering the question *when* is a continual process. As one learns more pathways and the complexities involved in the regulation of them all, the understanding of *when* (i.e., the integration of metabolism) is more challenging.

To study each metabolic pathway, use these six questions to design a "summary sheet" of the pathway. An example summary sheet is provided for glycolysis at the end of this chapter. The summary sheet should provide one with an overall knowledge of the pathway. Further details may be added, as indicated, for one's own understanding and/or the course in which one is enrolled.

KEY STRUCTURES TO BE ABLE TO DRAW AND RECOGNIZE

To learn structures of metabolic pathways, there are key structures one should know—both to be able to draw them and to recognize them. **Figure 2.1** indicates *some* of these key structures that will aide in identifying other structures of carbohydrate metabolism. These structures include glycerol, acetone, pyruvate, and the linear and cyclical forms of glucose and fructose. Note that the proper three-letter abbreviation for glucose is "**glc**," not

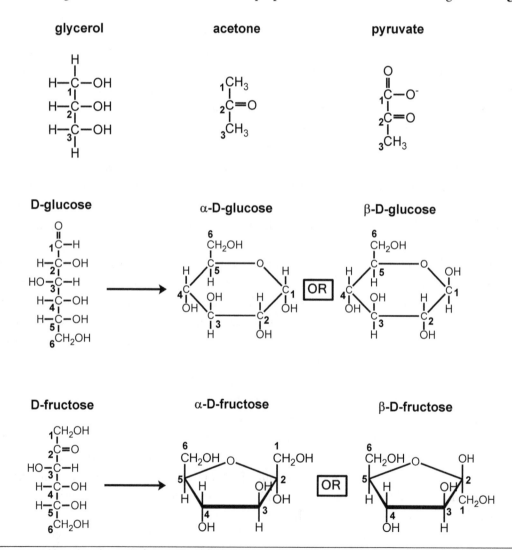

Figure 2.1: Key structures to know. Be able to draw and recognize them.

UNDERSTANDING BIOCHEMICAL PATHWAYS: A PATTERN-RECOGNITION APPROACH

"glu" (as many texts use). The three-letter abbreviation for the amino acid glutamate is "glu." Since it is confusing to use the same three-letter abbreviation for both molecules, this text will use the abbreviation "glc" for glucose.

For example, if one knows the structures of glucose and fructose, one does not need to memorize the structures for the first three steps of glycolysis. The names of the molecules for these first three steps are glucose, glucose 6-phosphate, fructose 6-phosphate, and fructose 1,6-bisphosphate. If one knows the structures of glucose and fructose and how the carbons are numbered, one knows where to put phosphate groups based on the names of the structures. Now it is okay to simply learn the order of the reactions, especially when one understands how the enzymes that carry out the reactions are named.

GLYCOLYSIS

As mentioned previously, every metabolic pathway covered in this text will answer four key questions that one should understand for a given pathway: *what, why, where,* and *how*. Glycolysis, as the first pathway covered, will begin to apply the principles that were introduced in the oxidation states chapter. Particular aspects of the regulation of glycolysis, for addressing the question *when*, will be covered to establish certain overarching concepts of metabolic regulation and to indicate some of the complexity involved in regulation of even a single pathway.

The overall reaction of aerobic glycolysis, shown in **figure 2.2**, catabolizes a molecule of glucose to yield two pyruvate, two ATP, two NADH, two protons, and two water molecules. For anaerobic glycolysis, the same net yield of ATP is produced, but there is no net NADH formed and two lactate are produced, rather than pyruvate. NADH is still formed at the same step, but it is utilized differently because the electrons carried on NADH cannot be sent to the mitochondria.

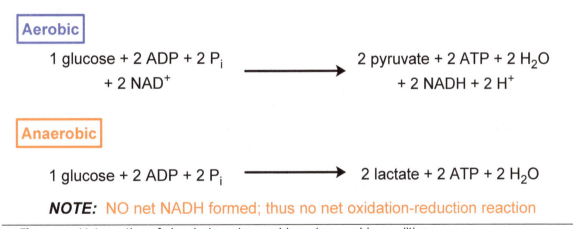

Figure 2.2: Net reaction of glycolysis under aerobic and anaerobic conditions.

To define the pathway of glycolysis (*what*): Glycolysis is the catabolism of glucose (a six-carbon molecule) to two pyruvates (three-carbon molecules) under aerobic conditions, or to two lactates (three-carbon molecules) under anaerobic conditions.

The purpose of glycolysis (*why*) is to produce ATP directly and reducing power in the form of NADH. Note that the structure of glucose has no phosphates on it and pyruvate has no phosphates on it. However, this pathway will generate ATP, which requires phosphorylation reactions. This pathway will put phosphates on intermediates and then use them to phosphorylate ADP to make ATP. None of the six carbons of glucose will be removed by the end of the glycolytic pathway. However, two out of the six carbons will be ready to be removed by creating two molecules of an α-keto acid. What happens to the NADH formed during glycolysis will differentiate aerobic versus anaerobic glycolysis.

Glycolysis occurs in the cytosol of the cell (*where*). Most catabolic processes occur in the mitochondria, as a purpose of catabolism is to generate ATP from the breakdown of nutrients. The reducing power generated goes to the electron transport chain for ATP production from oxidative phosphorylation. However, glycolysis is the only *catabolic* pathway that generates ATP from nutrient breakdown that does not depend on the mitochondria. (Note: ATP can certainly be made from de novo synthesis of nucleotides, but catabolic pathways produce chemical energy by phosphorylating a ribonucleoside diphosphate [i.e., ADP] to a ribonucleoside triphosphate [i.e., ATP].) This is why it is so important that the glycolytic pathway is not located in the mitochondria. If the electron transport chain is not working due to lack of oxygen, glycolysis becomes the only catabolic ATP-producing pathway via anaerobic glycolysis. As an example, mature red blood cells have no mitochondria. Mature red blood cells can only produce ATP by anaerobic glycolysis. The purpose of a red blood cell is to deliver oxygen to the tissues. Mitochondria are the organelles that use the most oxygen, as oxygen is the final electron acceptor of the electron transport chain. Therefore, it makes sense that a mature red blood cell does not have mitochondria, so it does not use what it is trying to deliver. Otherwise a red blood cell would preferentially use the oxygen itself.

The details of the individual reactions of glycolysis (*how*) will be covered in this chapter. Important key concepts to know include the starting and ending product structures, structures of key intermediates, the committed step reaction, and the enzymes and reactions that produce the end products. One should also know the regulated enzymes and what regulates them for the reactions that they carry out.

Figure 2.3 shows the entire glycolytic pathway with the names of the intermediates and enzymes, though the molecular structures are not shown. The ten reactions of glycolysis are numbered. The key to point out in this overview figure, which will be reiterated, is what happens to the molecules of NADH formed in the glyceraldehyde 3-phosphate dehydrogenase step. Under conditions where the mitochondria are functional (aerobic conditions), the cell shuttles the electrons carried by the hydride ion, into the mitochondria to be used to make ATP by the electron transport chain. Shuttles move the electrons carried by the hydride ion into the mitochondria and, in the process, regenerate cytosolic NAD^+ for the glyceraldehyde 3-phosphate dehydrogenase reaction.

Under anaerobic conditions the electron transport chain is not working. Under these conditions the electrons from this NADH cannot be shuttled into the mitochondria. However, glycolysis will stop as well if cells do not have a means of regenerating cytosolic NAD^+ for the glyceraldehyde 3-phosphate dehydrogenase step. Therefore, the purpose of the pyruvate to lactate conversion (see **figure 2.3**) by lactate dehydrogenase is the regeneration of cytosolic NAD⁺ under anaerobic conditions. The cell can continue to do glycolysis to produce ATP. Thus, the key difference between aerobic and anaerobic glycolysis is what happens to the NADH generated by the glyceraldehyde 3-phosphate dehydrogenase step.

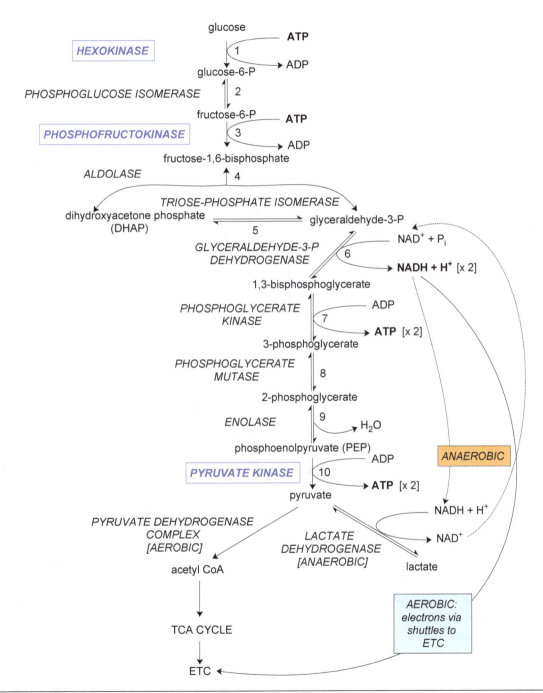

Figure 2.3: The glycolytic pathway.

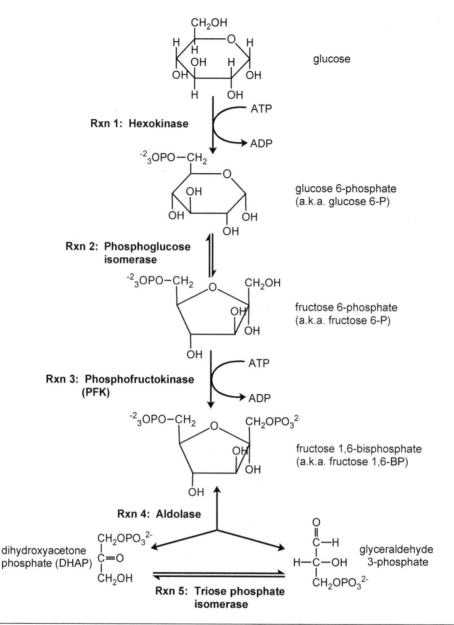

Figure 2.4: The first five steps of glycolysis.

Figure 2.4 shows the first five steps of glycolysis with the structures. Each of the reactions of glycolysis will be covered step-by-step, but this figure shows the first half of glycolysis drawn in sequence. In covering the details of the reactions, rationales will be presented to help one think about why the reactions must be carried out in a particular order. The reactions shown in the oxidation states flowchart (**figure 1.7**) present the pattern of the oxidation of carbon atoms, but there are other reaction types that need to occur in a cell. The objective in presenting these rationales is to help one remember the order that reactions need to occur, especially when the pathways include reactions not included in the oxidation states flowchart. The method for how the enzymes are named will also be covered.

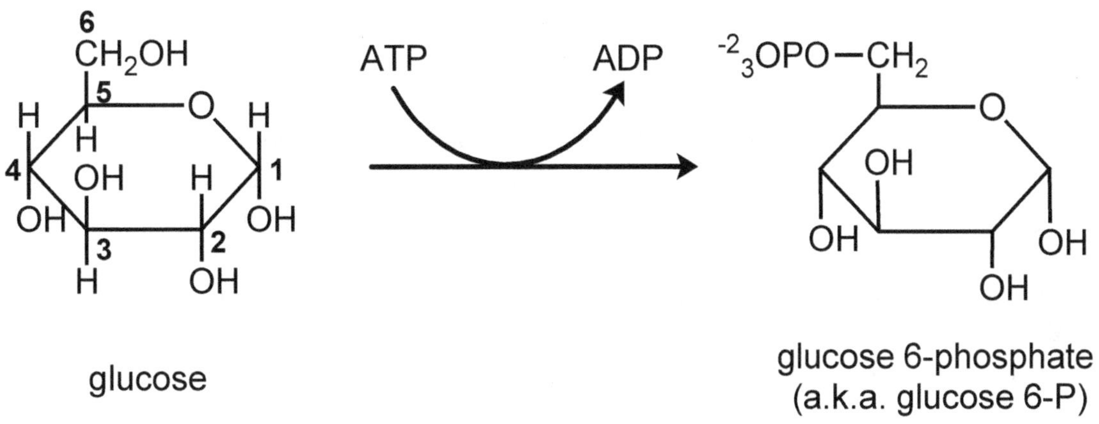

Enzyme: hexokinase (or glucokinase, an isozyme)

Figure 2.5: Reaction 1: Phosphorylation of glucose (glc) at carbon 6 using ATP.

The first reaction of glycolysis is shown in **figure 2.5**, which phosphorylates glucose to form glucose 6-phosphate. Glucose is shown with its six carbons numbered appropriately. If you compare and contrast glucose to glucose 6-phosphate, the reaction put a phosphate group on glucose at carbon 6. The phosphate group comes from ATP. This is a catabolic pathway, but to get it started the cell will need to use a little bit of ATP. The investment of some energy to start the pathway ensures the cell needs to do the pathway. Most catabolic pathways do require an input of energy in the beginning.

Kinases are enzymes that transfer a phosphate group to a molecule, usually from ATP. Kinases are named for the molecule that accepts the phosphate from ATP. Therefore, one would expect this enzyme to be called glucose kinase, or in short, *glucokinase*. Glucokinase is an isozyme form of the enzyme catalyzing this reaction in certain tissues. However, the enzyme that carries out this reaction in all cells is called *hexokinase*. Hexoses are six-carbon sugars. The suffix "–ose" is used in molecular names to indicate a sugar (a.k.a. carbohydrate). Hexokinase is an enzyme that can phosphorylate other six-carbon sugars such as fructose and mannose, although it has a high affinity for glucose.

Isozymes are different forms of an enzyme, encoded by different genes, that have the same general architecture and catalytic mechanism, but they are regulated differently. Hexokinase and glucokinase are isozymes. They both catalyze glucose + ATP to glucose 6-phosphate + ADP. Hexokinase is in all tissues of the body, while glucokinase is in the liver and islet cells of the pancreas. The K_m for hexokinase is low (0.1 mM) compared to the K_m for glucokinase (10 mM), which means hexokinase has a higher affinity for glucose than glucokinase. The V_{max} for hexokinase is also low relative to the V_{max} for glucokinase. The purpose for these differences is that the liver allows the glucose coming from the diet via portal circulation from the small intestine to pass through to ensure the blood glucose is maintained at about 5 mM. Once the blood glucose begins to rise past 5 mM, the hepatocytes (i.e., liver cells) begin to take up the excess glucose and quickly phosphorylate it to glucose 6-phosphate. Thus, the liver ensures that all other tissues have first access to glucose to use it or replenish their own glucose stores, if necessary.

Hexokinase is a regulated enzyme and is inhibited by high cellular concentrations of glucose 6-phosphate. Glucokinase is *not* inhibited by high cellular concentrations of glucose 6-phosphate. When cells have enough glucose, the glucose 6-phosphate concentration is high and inhibits hexokinase. Inhibition of hexokinase leads to inhibition of glucose uptake into the cell. Glucokinase cannot be inhibited by glucose 6-phosphate because the liver must take up all the excess glucose coming in via portal circulation to prevent blood glucose concentrations from becoming too high, leading to hyperglycemia. The liver cells can use the glucose to replenish their ATP concentration and glycogen stores. If there is still excess glucose coming in, the liver cells convert it to fat (a.k.a. triacylglycerols).

This first reaction of glycolysis is *irreversible*. An irreversible reaction is a reaction in which the Gibbs free energy of the reaction under cellular conditions (typically indicated as ΔG or $\Delta G'$) is equal to or more negative than –4 kcal/mol (–16.74 kJ/mol). In these cases the cell generally cannot manipulate the concentration of products to reactants in the equation for ΔG to make the reaction go in the reverse direction. Under cellular conditions the reaction catalyzed by hexokinase is even more favorable ($\Delta G \approx$ –8 kcal/mol) than under biochemical standard conditions ($\Delta G°'$ = –4 kcal/mol). One can think of the irreversibility of the hexokinase reaction in another manner. The Gibbs free energy under biochemical standard conditions ($\Delta G°'$) of the cleavage of a phosphoanhydride bond of ATP would yield approximately –7.3 kcal/mol. To reverse this reaction and phosphorylate ADP to make a phosphoanhydride bond of ATP, the cleavage of the phosphate group on carbon 6 would have to yield more than –7.3 kcal/mol (as some energy is always lost as heat). The $\Delta G°'$ of cleavage of an alcohol phosphate bond, such as this one, would only provide about –3 kcal/mol. Thus, an alcohol phosphate would not have enough energy in the bond to catalyze the reverse reaction, which would make a molecule of ATP.

The phosphorylation of glucose as glucose 6-phosphate traps the glucose inside the cell. When glucose enters the cell through specific glucose transport proteins, some of which are hormonally regulated, glucose immediately gets phosphorylated and cannot go back across the cellular membrane. Cells take up glucose molecules for their "own" use. Only particular cells of the liver and kidneys have the enzyme glucose 6-phosphatase that can clip off the phosphate group from glucose 6-phosphate to form glucose, which can be released back into the blood (as will be covered in the pathway of gluconeogenesis).

The reaction catalyzed by hexokinase is *not* the committed step of glycolysis. A committed step is the reaction of a pathway in which the product of that reaction is restricted to proceeding on through the pathway. Glucose 6-phosphate, though, is an intermediate for other pathways. Glucose 6-phosphate is a key branch point between several pathways, such as glycogen synthesis, glycogen breakdown, the pentose phosphate pathway, or the formation of glucose from gluconeogenesis in specific tissues (i.e., the liver and kidneys).

There is a particular rationale one can think of to understand the sequence of the first three steps of glycolysis. For these first steps of glycolysis, the goal is to phosphorylate the two end carbons of glucose (1 and 6), but only if these carbons have a primary alcohol group. This first reaction does phosphorylate carbon 6, which has a primary alcohol. Carbon 1, the other end carbon of glucose, does not have a primary alcohol. This is more obvious when glucose is drawn in its straight chain form (see **figure 2.6**), in which one can see that carbon 1 of glucose is an aldehyde. The cell will phosphorylate it, but as a primary alcohol. Recall from the oxidation states flowchart (**figure 1.7**) that an aldehyde can be isomerized to a ketone, and both are at the same oxidation state. If one knows the structure of glucose (an aldose or aldehyde sugar) and fructose (a ketose or keto sugar), one can draw this reaction, which is shown in **figure 2.6**. Notice that in the molecule of fructose 6-phosphate, carbon 1 now has a primary alcohol. This is the purpose of reaction 2 of glycolysis: the isomerization of glucose 6-phosphate to fructose 6-phosphate to produce carbon 1 with a primary alcohol functionality. This primary alcohol will then be phosphorylated in reaction 3 of glycolysis.

Reaction 2 isomerizes glucose 6-phosphate to fructose 6-phosphate, and both the cyclical and the linear structures are shown in **figure 2.6**. Both molecules still have a phosphate group on carbon 6. In the cyclical structures, glucose 6-phosphate has a pyranose ring (a six-membered ring). Fructose 6-phosphate has a furanose ring (a five-membered ring). The isomerization of these two molecules is readily reversible. This is

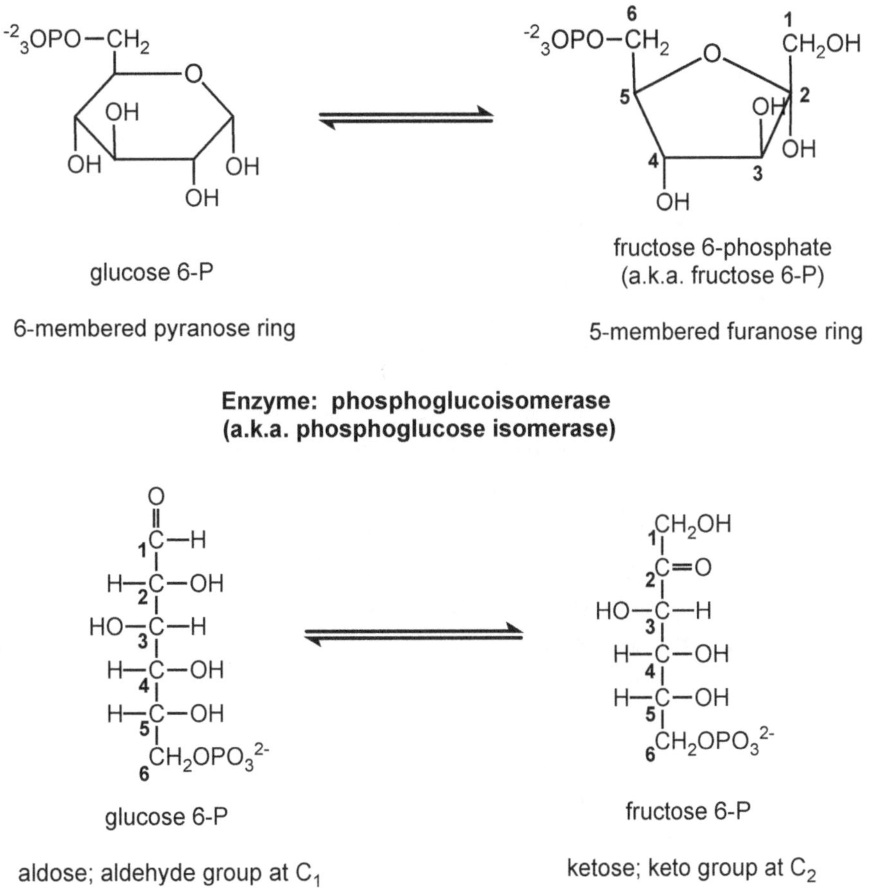

Figure 2.6: Reaction 2: Isomerization of glucose 6-phosphate to fructose 6-phosphate. Both shown as cyclical and linear forms of the sugars.

an isomerization reaction, and thus the enzyme that catalyzes it is an *isomerase*. There are no definitive rules for naming isomerases. In this case the enzyme is named after the substrate. One could call the enzyme glucose 6-phosphate isomerase, but the enzyme name drops the "6" and puts the "phospho" in front to call it *phosphoglucose isomerase*, or more often simply known as *phosphoglucoisomerase*.

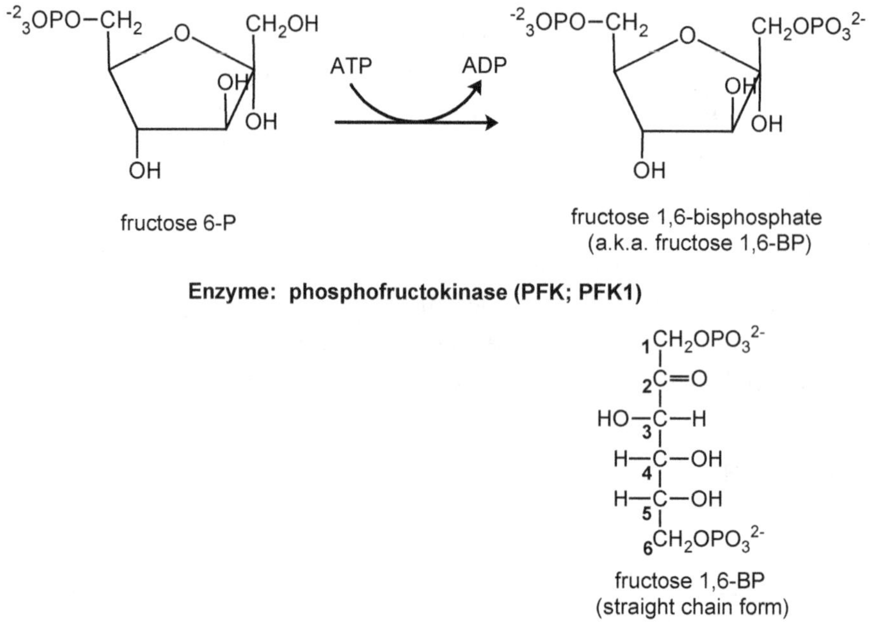

Figure 2.7: Reaction 3: Second phosphorylation step using ATP. (Note: Fructose 1,6-bisphosphate is shown as both cyclical and linear structures.)

Now that fructose 6-phosphate has a primary alcohol on carbon 1, this hydroxy group can be phosphorylated in reaction 3, as shown in **figure 2.7**. ATP is the donor of the phosphate group. As one would expect, the name of the product is fructose 1,6-bisphosphate. The name of the molecule indicates its structure, as long as one knows the structure of fructose. Enzymes that phosphorylate using ATP are called *kinases*. Kinases are named for the molecule accepting the phosphate from ATP, which is fructose 6-phosphate. One could call the enzyme fructose 6-phosphate kinase. However, the name of the enzyme drops the "6" and puts the "phospho" at the beginning of the enzyme name to call it *phosphofructokinase*. (Notice the enzyme name is *not* "bisphosphofructokinase," which would indicate the name for the product of the reaction.) The enzyme is commonly abbreviated as PFK, and to be very specific, PFK1 for *phosphofructokinase 1*. There is a phosphofructokinase 2 (PFK2) that is involved in the regulation of this step, which will be discussed later regarding the regulation of glycolysis.

Phosphofructokinase catalyzes another *irreversible* step of glycolysis. There are four kinase reactions in glycolysis, and three out of the four kinase reactions are irreversible. The ΔG of this reaction is about −5.0 kcal/mol, so the cell cannot manipulate the concentrations of reactants and products to allow the reaction to run in the reverse direction. Again, the cleavage of the alcohol phosphate bond on carbon 1 of fructose 1,6-bisphosphate does not have enough energy to create a phosphoanhydride bond of ATP for the reverse reaction. This third reaction of glycolysis is the committed step of glycolysis, and it is highly regulated. Once the cell makes fructose 1,6-bisphosphate using PFK1, this molecule will continue through glycolysis.

Recall that one of the goals of glycolysis is to make ATP. The pathway starts with glucose, which has no phosphates, and the product of the pathway, pyruvate, also has no phosphates. The goal, then, is to take the phosphates that are being put on this molecule and use them to create more ATP. How will the cell achieve a net yield of ATP when it has used two ATP molecules to put on the two phosphate groups, so far? Another reaction of glycolysis will attach an additional phosphate group to a particular intermediate that does not require the use of ATP.

Now the cell has phosphorylated the end carbons, as was the goal of these first three steps. The next enzyme will cut fructose 1,6-bisphosphate in half for reaction 4, forming two three-carbon molecules, each containing a phosphate group, as shown in **figure 2.8**. **Figure 2.8** shows fructose 1,6-bisphosphate in its linear form (rather than its normal cyclical form) to indicate how the molecule is divided. This figure, though, does not imply anything regarding the actual enzymatic mechanism. The molecular names of the two products indicate their structures, if one knows two of the key structures shown in **figure 2.1**: acetone and glycerol. The structure of acetone was the ketone molecule that was also drawn in the oxidation states flowchart (**figure 1.7**). Glycerol is another simple molecule to draw, since it has three carbons with a hydroxy group on each carbon. A carbon makes four bonds, so fill in the remaining carbon bonds with hydrogens to complete the structure of glycerol. If one can draw those two molecules, the remaining glycolytic intermediates (except for the last reaction) can be drawn based on the names of the structures.

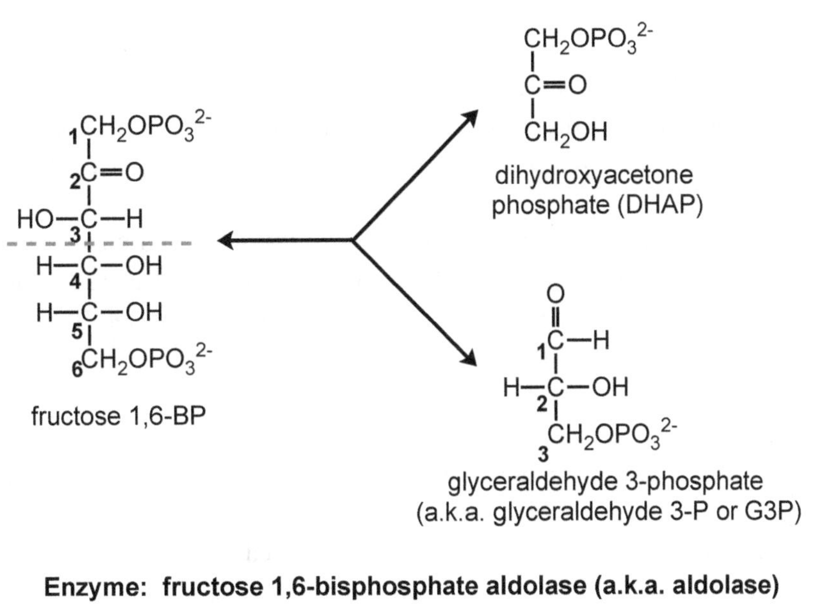

Enzyme: fructose 1,6-bisphosphate aldolase (a.k.a. aldolase)

Figure 2.8: Reaction 4: Cleavage of fructose 1,6-bisphosphate into two three-carbon phosphorylated molecules.

METABOLISM OVERVIEW AND GLYCOLYSIS

When fructose 1,6-bisphosphate is cut in half, the products of the reaction are two three-carbon molecules. One of the products is a ketone, dihydroxyacetone phosphate (DHAP), and the other product is an aldehyde, glyceraldehyde 3-phosphate. For dihydroxyacetone phosphate, first draw acetone but replace one hydrogen on each end carbon with a hydroxy group—i.e., dihydroxyacetone. Since this is a symmetrical molecule, pick one of the hydroxy groups and put a phosphate group on it to complete the structure for dihydroxyacetone phosphate. In **figure 2.8**, the phosphate is put on the end carbon that matches how fructose 1,6-bisphosphate would be drawn cut in half to produce the ketone. It is best to learn the entire name of the molecule, not simply the abbreviation, DHAP. Dihydroxyacetone phosphate, the full name of the molecule, indicates its molecular structure.

Glyceraldehyde 3-phosphate is the aldehyde product of this reaction. For this molecule, first draw glycerol, but draw an aldehyde functionality on one of the end carbons (instead of the hydroxy group and hydrogens) forming glyceraldehyde. The aldehyde group is the highest oxidation state of this molecule, which makes this carbon 1. The name of the molecule is glyceraldehyde 3-phosphate, which means the phosphate group is drawn on the hydroxy group of carbon 3—the other end carbon. Now the two products of this reaction have been drawn, dihydroxyacetone phosphate and glyceraldehyde 3-phosphate.

This reaction is very endergonic under standard $\Delta G°'$ conditions (about +5.7 kcal/mol), but in the cell the ΔG is only about −0.5 kcal/mole, which makes the reaction readily reversible. The enzyme for reaction 4 is named for a classic organic reaction called an aldol cleavage. An aldol cleavage is a reaction in which a molecule is cleaved to form the products of an aldehyde and a ketone. An aldol condensation, which is the reverse reaction, is a reaction that combines an aldehyde and a ketone to form a larger product. Thus, the enzyme name for this reaction is *fructose 1,6-bisphosphate aldolase*, more commonly known simply as *aldolase*.

Now consider the two products of reaction 4. Dihydroxyacetone phosphate has a keto group, which is a functional group that cannot be further oxidized. Glyceraldehyde 3-phosphate, though, has an aldehyde group that can be further oxidized to a carboxylic acid—a goal of catabolic processes. *Therefore, glyceraldehyde 3-phosphate is the molecule that can proceed through the rest of glycolysis.* A keto group, while it cannot be further oxidized to a carboxylic acid, can be isomerized to an aldehyde—and then oxidized to a carboxylic acid. Compare the molecules of dihydroxyacetone phosphate and glyceraldehyde 3-phosphate. Both molecules have the same number of carbons, oxygens, hydrogens, and phosphates. These two molecules are isomers of one another.

Reaction 5, shown in **figure 2.9**, isomerizes the dihydroxyacetone phosphate (the ketone) to glyceraldehyde 3-phosphate, which are at the same oxidation state. The enzyme that catalyzes this reaction is therefore called an *isomerase*. Again, there are no standard naming conventions for isomerases. These two molecules are, in general terms, three-carbon phosphorylated sugars—i.e., triose phosphates. Thus, the enzyme name is *triose phosphate isomerase*. This reaction is rapid and readily reversible. A pool of both molecules is needed to carry out reactions in the cell. At equilibrium there is actually more dihydroxyacetone phosphate than glyceraldehyde 3-phosphate. Remember, though, how metabolism

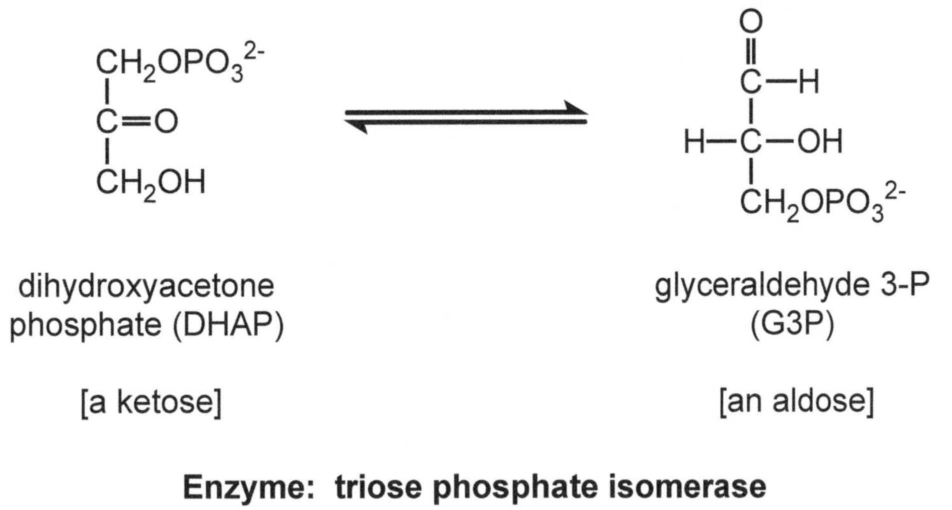

$$\begin{array}{c} CH_2OPO_3^{2-} \\ | \\ C=O \\ | \\ CH_2OH \end{array} \quad \rightleftharpoons \quad \begin{array}{c} O \\ \| \\ C-H \\ | \\ H-C-OH \\ | \\ CH_2OPO_3^{2-} \end{array}$$

dihydroxyacetone phosphate (DHAP)   glyceraldehyde 3-P (G3P)

[a ketose]   [an aldose]

Enzyme: triose phosphate isomerase

Figure 2.9: Reaction 5: Isomerization of dihydroxyacetone phosphate to glyceraldehyde 3-phosphate.

works—maintaining a particular ratio of products to reactants will allow cellular reactions to go in the direction necessary. Keeping the concentration of products low and the concentration of the reactants high will cause the reaction to go in the direction needed. Using the glyceraldehyde 3-phosphate, as soon as it is formed, in the next reaction forces more dihydroxyacetone phosphate to be isomerized to glyceraldehyde 3-phosphate.

At this point, for the purposes of ultimately calculating the maximum yield of ATP from the complete catabolism of a molecule of glucose, there are now two molecules of glyceraldehyde 3-phosphate that will continue through the remainder of glycolysis. The rest of the reactions of glycolysis will then be carried out two times.

Figure 2.10 shows the last five reactions of glycolysis with all the structures. Refer to this figure to review how these last five reactions are carried out in sequence. As the last half of glycolysis is covered, one needs to keep in mind the goals of glycolysis: The pathway is working to oxidize carbons to carboxylic acids, as well as set them up to be good leaving groups as CO_2, and to produce ATP and reducing power in the form of NADH.

The sequence shown for reaction 6 in **figure 2.11** is not mechanistically correct. The actual enzyme mechanism does not produce the intermediate shown, which is 3-phosphoglycerate. 3-Phosphoglycerate is an intermediate of glycolysis, but it is not an actual intermediate of reaction 6. The true reaction intermediate is attached to a sulfhydryl group of the enzyme, ultimately forming a thioester. The simplified reaction sequence shown is based on the patterns indicated in the oxidation states flowchart (**figure 1.7**). In reaction 6 the simplified scheme shown in **figure 2.11** indicates that the enzyme carries out two steps. In the first step the aldehyde group of glyceraldehyde 3-phosphate will be oxidized to a carboxylic acid.

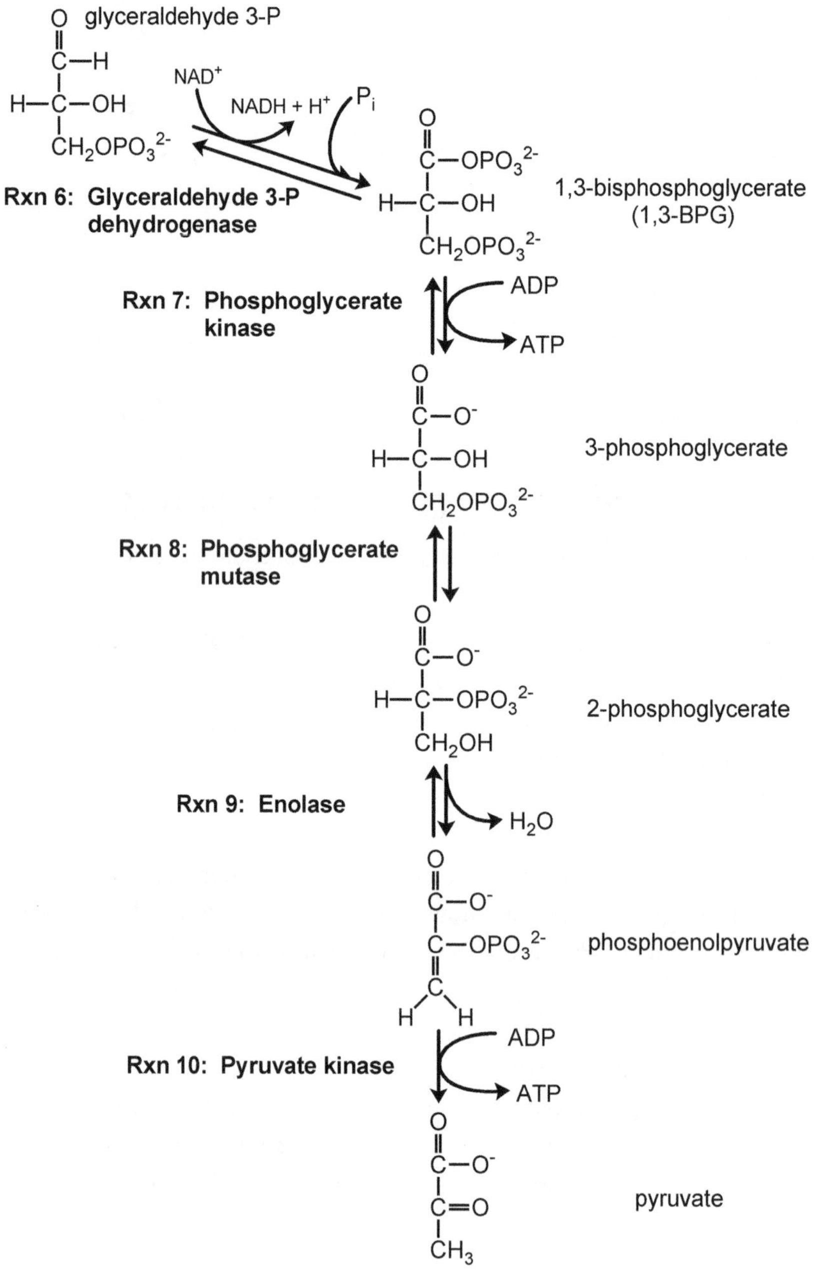

Figure 2.10: The last five steps of glycolysis.

The next step is to add a phosphate group to the carboxylic acid group using inorganic phosphate (P_i), rather than ATP as a donor.

In the first step of reaction 6, the molecule of water needed to go from an aldehyde to a carboxylic acid (as shown in the oxidation states flowchart, **figure 1.7**) is shown in **figure 2.11**. Recall, though, that only part of the water molecule is added, which is why this is an oxidation-reduction reaction and not a hydration

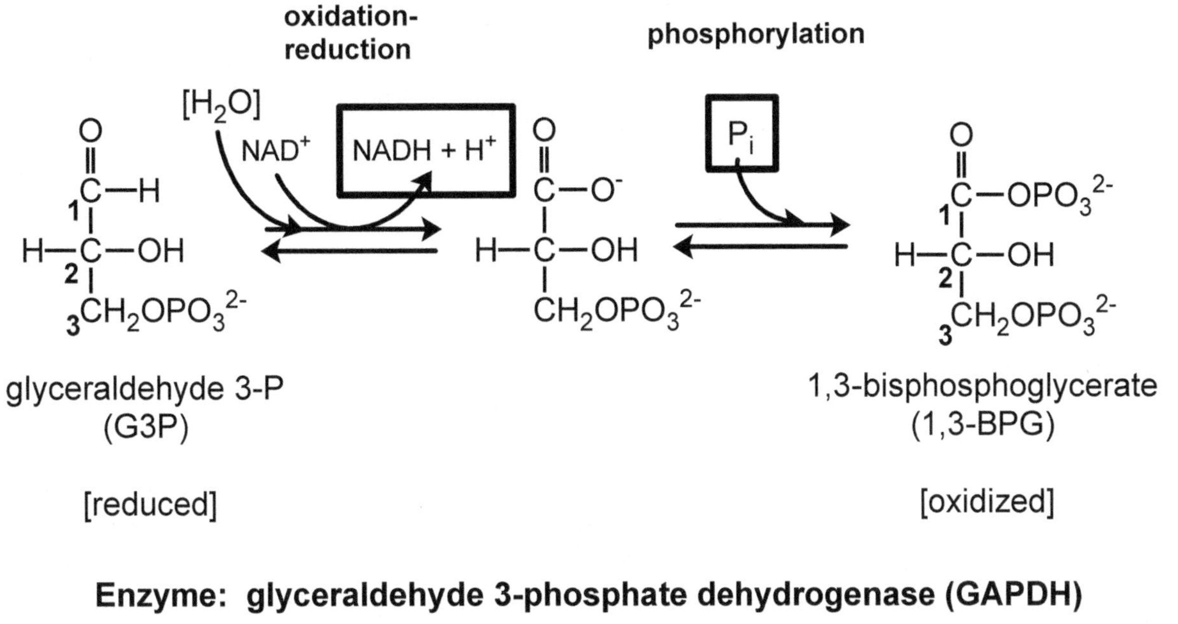

Figure 2.11: Reaction 6: Oxidation and phosphorylation of glyceraldehyde 3-phosphate.

Enzyme: glyceraldehyde 3-phosphate dehydrogenase (GAPDH)

reaction. One of the hydrogens is lost from the aldehyde group, and one of the hydrogens is lost from the water molecule. Oxidation and reduction reactions are always coupled. As mentioned in chapter 1, FAD is used when oxidizing an alkane to an alkene. All other oxidations use NAD^+. In this reaction an aldehyde is oxidized to a carboxylic acid; hence, NAD^+ is reduced to $NADH + H^+$.

The second step is the phosphorylation of the –OH group (or $–O^-$, if deprotonated as shown in **figure 2.11**) of the carboxylic acid. However, the cell does not use ATP for this reaction because the goal is to get a net production of ATP. Two ATP were used to put the phosphate groups on the two molecules of glyceraldehyde 3-phosphate created. Inorganic phosphate (P_i, or HPO_4^{2-}) is now used to phosphorylate the carboxylic acid group—not ATP.

The final product of this enzymatic reaction is 1,3-bisphosphoglycerate. The name indicates the structure of the molecule. The "-ate" ending indicates a deprotonated carboxylic acid group (i.e., carboxylate ion) on the base structure of glycerol. Draw glycerol, but draw a deprotonated carboxylic acid group on an end carbon (instead of the –OH group and hydrogens). This molecule would be called glycerate. A carboxylic acid group is the highest oxidation state, so that carbon becomes carbon 1 of the molecule. The name of the molecule indicates that there are two phosphate groups ("bisphospho"). One of the phosphate groups is on carbon 1, the other is on carbon 3—as shown in **figure 2.11**. At this point, there are now two molecules of 1,3-bisphosphoglycerate because this reaction will be done two times for the two molecules of glyceraldehyde 3-phosphate. With two molecules of 1,3-bisphosphoglycerate formed from a single glucose molecule, there are now four phosphate groups that will be used to create four molecules of ATP (for a net yield of two ATP) by the end of the glycolytic pathway—which is not a lot, but often adequate when oxygen is lacking.

For naming the enzyme that carries out this two-step reaction, anytime an enzyme carries out multiple reactions and one of them is a redox reaction using NAD^+ or FAD, the enzyme is always called a *dehydrogenase*. Dehydrogenases are always named for the *more reduced* molecule, which is glyceraldehyde 3-phosphate for this reaction. This enzyme is then named *glyceraldehyde 3-phosphate dehydrogenase*, which is often abbreviated as GAPDH (pronounced "gap D H"). The full name, though, indicates the reaction. It is recommended to use the abbreviation only after one has learned the full enzyme name.

This is the only oxidation-reduction step in the pathway of aerobic glycolysis, and it couples oxidation with phosphorylation. This reaction is readily reversible. What happens to the NADH formed in this reaction (i.e., how the cell regenerates cytosolic NAD^+) is the difference between aerobic and anaerobic glycolysis. The cell must have a pool of cytosolic NAD^+ (the oxidized form) for this reaction to be carried out.

Another key point is that there are now four phosphate groups (two phosphates per molecule of 1,3-bisphosphoglycerate) to ultimately be removed for the formation of four ATP molecules from the catabolism of a single glucose molecule. However, these phosphate groups must be good leaving groups and the bonds cleaved must have more than enough energy to create a phosphoanhydride bond of ATP that has inherently about –7.3 kcal/mol. (Note that some energy is always lost as heat.) The phosphate group attached to the carboxylic acid group of carbon 1 has a $\Delta G^{o\prime}$ of about –11.5 kcal/mol, which is more than enough energy to create a bond that has –7.3 kcal/mol. The alcohol phosphate on carbon 3 only has a $\Delta G^{o\prime}$ of about –3 kcal/mol. The phosphate group on carbon 1 will leave first, and the remaining steps of glycolysis will need to elevate the leaving potential of the phosphate group on carbon 3.

Reaction 7, as shown in **figure 2.12**, produces the first ATP. This reaction is where the cell "breaks even" in its yield of ATP from glycolysis. Two molecules of 1,3-bisphosphoglycerate go through this reaction to produce two ATP, but the pathway used two ATP in the first half. The phosphate group from carbon 1 of 1,3-bisphosphoglycerate is transferred to ADP to make ATP, forming the product 3-phosphoglycerate, whose name indicates its structure. The "–ate" indicates that the molecule has a deprotonated carboxylic

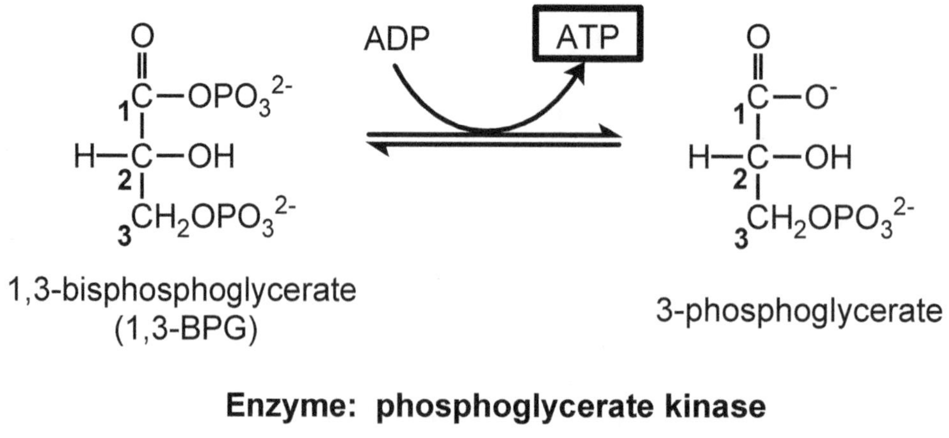

Enzyme: phosphoglycerate kinase

Figure 2.12: Reaction 7: Formation of the first ATP.

acid group on the "base molecule" glycerol—forming *glycerate*. The carboxylic acid carbon is carbon 1, as the highest oxidation state. Thus, the phosphate group is on the hydroxy group of the other end carbon, carbon 3.

This reaction is carried out by a kinase, though it may not look like it because a kinase is an enzyme that transfers a phosphate from ATP to an acceptor. This reaction produces ATP. Nevertheless, the enzyme that catalyzes reaction 7 and the enzyme that catalyzes reaction 10 are kinases. These two kinases are named for the reverse reactions. Naming the enzyme for the reverse reaction is the correct way to name the kinase—as if ATP is donating a phosphate group to an acceptor molecule, 3-phosphoglycerate. The enzyme is called *phosphoglycerate kinase*, as the enzyme name does not include the "3." The reaction catalyzed by phosphoglycerate kinase is the only kinase reaction of the four kinase reactions in glycolysis that is actually reversible.

Reaction 7 is termed a substrate-level phosphorylation reaction because it transfers a high-energy phosphate leaving group from an organic molecule to ADP to make ATP. In contrast, oxidative phosphorylation is carried out by the electron transport chain. Oxidative phosphorylation couples the energy created from a proton gradient, which is produced by the movement of electrons, to the phosphorylation of ADP to make ATP. There are not many enzymes that can create a phosphoanhydride bond of ATP via a substrate-level phosphorylation reaction.

Now consider what needs to happen in the last three steps of glycolysis—reactions 8, 9, and 10—which includes the application of concepts covered in chapter 1. At this point in glycolysis, two of the six original carbons of a glucose molecule are now at the level of a carboxylic acid (i.e., two molecules of 3-phosphoglycerate)—one of the goals of catabolic processes. Carboxylic acid groups are stable, though, and are not good leaving groups. In biological systems a keto group must be placed alpha or beta to the carboxylic acid carbon to make the carboxylic acid group a good leaving group. However, 3-phosphoglycerate is a three-carbon molecule, and a keto group is a carbonyl group attached to two other carbons. Therefore, the only place to put a keto group is on carbon 2 of the molecule, creating an α-keto acid. Decarboxylation of an α-keto acid is more difficult enzymatically, but it can be done. The decarboxylation of pyruvate, an α-keto acid, to form CO_2 does not occur in glycolysis. Two molecules of pyruvate are formed at the end of glycolysis, so all six carbons of the original glucose molecule are still in the cytosol of the cell. The final product, pyruvate, also does not have any phosphate groups. Another goal is to make the phosphate group on 3-phosphoglycerate a good enough leaving group that it can be used to phosphorylate ADP to make ATP. Thus, the two goals that must be accomplished by reactions 8, 9, and 10 are to elevate the leaving potential of the phosphate group and create a keto group on carbon 2.

In reaction 8, shown in **figure 2.13**, the first step toward accomplishing the two goals is to move the phosphate group from the hydroxy group of carbon 3 to the hydroxy group of carbon 2. The names indicate the structures: 3-phosphoglycerate and 2-phosphoglycerate. Drawing glycerate has already been described; now just draw the phosphate group on the hydroxy group of carbon 2. This is an isomerization reaction and is readily reversible. The number of carbons, hydrogens, oxygens, and phosphates are the same on both molecules, just arranged differently. One could call the enzyme for this reaction an isomerase,

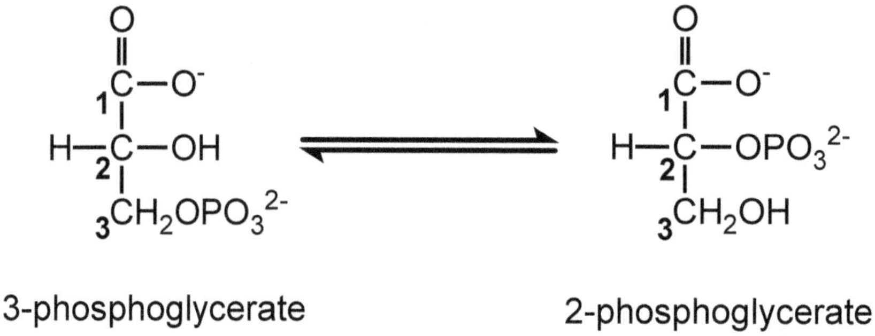

$$\begin{array}{c}\text{O}\\\parallel\\{}_1\text{C}-\text{O}^-\\\mid\\\text{H}-{}_2\text{C}-\text{OH}\\\mid\\{}_3\text{CH}_2\text{OPO}_3^{2-}\end{array}\quad\rightleftharpoons\quad\begin{array}{c}\text{O}\\\parallel\\{}_1\text{C}-\text{O}^-\\\mid\\\text{H}-{}_2\text{C}-\text{OPO}_3^{2-}\\\mid\\{}_3\text{CH}_2\text{OH}\end{array}$$

3-phosphoglycerate → 2-phosphoglycerate

Enzyme: phosphoglycerate mutase

Figure 2.13: Reaction 8: Rearrangement of the phosphate group.

but this reaction specifically moved a functional group (i.e., a phosphate group). Enzymes that catalyze intramolecular shifts of functional groups on a molecule (i.e., phosphate groups, amino groups, etc.) are a special class of isomerases called *mutases*. Therefore, the name of this enzyme drops the numbers from the molecule names ("2" and "3") and is called *phosphoglycerate mutase*.

It does not seem like this reaction accomplished much, because that phosphate group is still an alcohol phosphate group, which is a poor leaving group. However, the two goals are that the phosphate group can be used to form ATP, and a keto group needs to be created on carbon 2. When the phosphate group leaves from carbon 2, a keto group will be created on carbon 2. This is the rationale one can use to explain why the phosphate group is moved to carbon 2. First the phosphate group is moved, then the next reaction elevates its leaving potential.

Reaction 9, shown in **figure 2.14**, forms an enol group, specifically an enol phosphate. This is a dehydration reaction, which is a loss of water from the molecule (i.e., the hydrogen from carbon 2 and the hydroxy

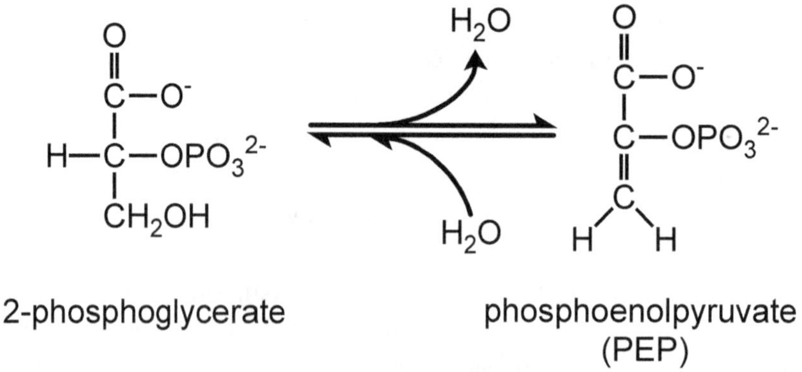

2-phosphoglycerate → phosphoenolpyruvate (PEP)

Enzyme: enolase

Figure 2.14: Reaction 9: Formation of an enol phosphate.

UNDERSTANDING BIOCHEMICAL PATHWAYS: A PATTERN-RECOGNITION APPROACH

group from carbon 3). As shown in the oxidation states flowchart (**figure 1.7**), dehydration of a molecule yields a molecule with a double bond (an alkene). This dehydration reaction is reversible.

The dehydration of 2-phosphoglycerate yields phosphoenolpyruvate (PEP), a product that has a double bond between carbons 2 and 3. The name of the product, phosphoenolpyruvate, does not explicitly indicate its structure. One needs to know the structure of the three-carbon molecule pyruvate, formed in reaction 10, to draw the structure based on its name. Phosphoenolpyruvate is three carbons, and the name indicates it has a deprotonated carboxylic acid group (i.e., the "-ate" ending) and an enol phosphate. The "ene" part of "enol" indicates the double bond (i.e., alkene) functional group. The "-ol" of "enol" indicates an alcohol group (hydroxy group), which is still on carbon 2 because the hydroxy group was removed from carbon 3 in this reaction. The phosphate is also still on the hydroxy group of carbon 2, as it was for the substrate 2-phosphoglycerate—forming the "enol phosphate."

Most enzymes that carry out dehydration reactions are called *dehydratases* and are named for the molecule that loses the water molecule. Most enzymes that carry out hydration reactions are called *hydratases* and are named after the molecule that gains the water molecule. However, many dehydration and hydration reactions are reversible, so some enzyme names for these types of reactions do not follow the naming rules. The enzyme name for this reaction is an exception to the naming rules. The enzyme that catalyzes reaction 9 is simply called *enolase*, for the enol functional group it creates on phosphoenolpyruvate.

The formation of the enol phosphate by this dehydration reaction achieves the purpose of markedly elevating the leaving group potential of the phosphate group. The ΔG°' of the enol phosphate bond is now about −15 kcal/mol, which is more than enough energy to make a phosphoanhydride bond of ATP. Now this phosphate group can be used to phosphorylate ADP to make ATP. Also, when the phosphate group leaves from carbon 2, a keto group will be formed on that carbon.

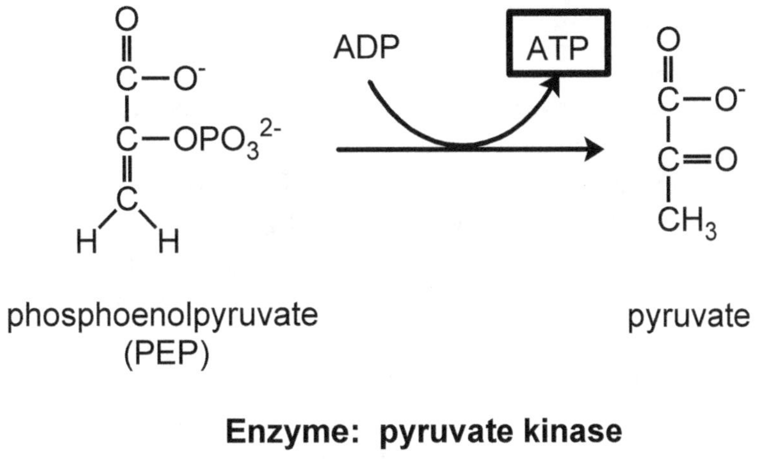

Figure 2.15: Reaction 10: Formation of pyruvate and the second ATP.

Figure 2.15 shows reaction 10, the last reaction of glycolysis (at least for aerobic glycolysis). This reaction provides the net yield of two ATP for the pathway in the formation of two molecules of pyruvate. In this reaction, the phosphate group on carbon 2 of phosphoenolpyruvate is transferred to ADP to make ATP, resulting in the formation of pyruvate. This reaction is also a substrate-level phosphorylation reaction.

The name "pyruvate" does not indicate much about its structure. The "–ate" ending indicates there is a deprotonated carboxylic acid group (which must be on an end carbon that becomes carbon 1), but this is a structure one must know (like glucose and fructose). Pyruvate is a three-carbon α-keto acid. Since it is three carbons, the keto group must be on the middle carbon. Therefore, draw the keto group on carbon 2, and then draw a methyl group for carbon 3.

As explained for reaction 7, this enzyme is also going to be called a kinase, but it is named for the reverse reaction. However, this reaction is *irreversible*. This reaction does not actually go "in reverse." Kinases, though, are always named as if ATP is donating a phosphate group to an acceptor molecule, pyruvate in this case. Therefore, this enzyme is called *pyruvate kinase* (*not* phosphoenolpyruvate kinase).

As mentioned, this reaction is irreversible, with a ΔG of about –5.5 kcal/mol, and it is regulated. Of the four steps in glycolysis that use kinases, three are irreversible. These three irreversible steps are the regulated steps of glycolysis.

In this last step of glycolysis, an α-keto acid has been formed. There is no loss of carbons in glycolysis, but the two carboxylic acid groups (i.e., two molecules of pyruvate) have been set up as good leaving groups because of the keto group on the alpha carbon of pyruvate. This catabolic pathway has also provided a net yield of two ATP. The fate of pyruvate is going to depend on whether molecular oxygen is available, whether pyruvate needs to be further catabolized, or whether certain amino acids are needed. Recall that pyruvate can be converted to alanine through a transaminase reaction.

Regulation of glycolysis

Another key question one should strive to understand about any metabolic pathway is: *When*? When (i.e., under what conditions) is this pathway activated ("on") or inhibited ("off")? With which other pathways does the cell run the pathway simultaneously or in sequence? Which pathways cannot be on simultaneously in the same cell (i.e., reciprocal pathways like glycolysis and gluconeogenesis)? The ability to answer the question *when* helps one achieve understanding of how metabolic pathways are integrated. Two basic concepts of metabolic pathway regulation will be stressed when covering a given metabolic pathway in this text:

1. How does the "energy state" of the cell, meaning the ratio of cellular concentrations of ATP to ADP (or AMP), affect the activation or inhibition of the pathway?
2. When two enzymes that use the same reactants (i.e., substrates and/or products) are in the same cellular compartment, how does the cell regulate which enzyme is active versus inhibited?

Therefore, before discussing the fates of the products of glycolysis, a review of some of the important regulators of glycolysis are presented. Regulation of metabolic pathways is very complex and includes levels of regulation from the expression of the genes encoding the enzymes, the activities of the expressed enzymes, the accessibility of the substrates and products, as well as the degradation of the enzymes. The glycolytic pathway will be used to present a brief insight into this complexity, but only at the level of activity of the enzymes.

There are three irreversible kinase reactions in glycolysis catalyzed by hexokinase, phosphofructokinase 1 (PFK1), and pyruvate kinase. These three reactions have to be bypassed in the reciprocal pathway of gluconeogenesis, which is the de novo synthesis of a molecule of glucose from two molecules of pyruvate. Gluconeogenesis will be covered in chapter 5. However, the enzymes that bypass the irreversible reactions of glycolysis are introduced here for the discussion of reciprocal regulation of these two pathways.

The reactions catalyzed by hexokinase and PFK1 are bypassed in gluconeogenesis using enzymes called *phosphatases*. A phosphatase is an enzyme that clips a phosphate group off of a molecule and releases it as inorganic phosphate (P_i). To bypass the hexokinase reaction, *glucose 6-phosphatase* (named for the molecule which loses the phosphate group) is the enzyme that catalyzes the reciprocal reaction in gluconeogenesis. Only two organs in your body have that enzyme, the liver and kidneys, because these organs are responsible for maintaining the blood glucose supply. For the PFK1 reaction, *fructose 1,6-bisphosphatase* (FBP1) catalyzes the reciprocal reaction in gluconeogenesis (shown in **figure 2.16**; note the full name of FBP1 is written in the lower right corner of the figure). Fructose 1,6-bisphosphatase is named for the molecule that loses the phosphate group. In gluconeogenesis, bypassing the pyruvate kinase reaction of glycolysis requires two enzymes and a transporter across the inner mitochondrial membrane. These gluconeogenic enzymes are *pyruvate carboxylase* and *phosphoenolpyruvate carboxykinase* (PEPCK). Glycolysis and gluconeogenesis cannot be active (on) simultaneously in the same cell because it would create a futile cycle, in which glucose is broken down by glycolysis only to be resynthesized by gluconeogenesis. Running these pathways simultaneously in the same cell would yield no net energy, and no chemical or biological work would be done.

Hexokinase is the first regulated step of glycolysis. Recall that hexokinase catalyzes reaction 1, the phosphorylation of glucose by ATP to form glucose 6-phosphate and ADP. Glucose 6-phosphate is an allosteric inhibitor of hexokinase. There is a separate binding site on the enzyme hexokinase for glucose 6-phosphate. As glucose 6-phosphate builds up in a cell, glucose 6-phosphate binds to this allosteric site and inhibits this enzyme. This inhibition helps control glucose uptake into the cell. Recall that glucokinase, the liver isozyme, is not inhibited by glucose 6-phosphate. The hexokinase reaction is not the key regulated step, because glucose 6-phosphate does not have to go through glycolysis.

The reaction catalyzed by phosphofructokinase 1 is the committed step of glycolysis and the primary regulatory control point of the pathway. PFK1 catalyzes the phosphorylation of fructose 6-phosphate by ATP to form fructose 1,6-bisphosphate and ADP, as shown in **figure 2.16**. Phosphofructokinase 1 has several regulatory sites on the enzyme for the binding of several allosteric activators and inhibitors. Allosteric activators of PFK1 include fructose **2,6-bisphosphate** (shown in **figure 2.16** and will be covered in detail)

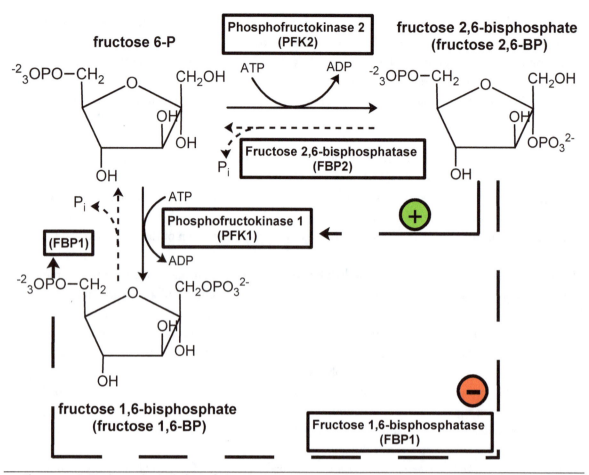

Figure 2.16: Reciprocal regulation of glycolysis and gluconeogenesis via fructose 2,6-bisphosphate (fructose 2,6-BP).

Figure 2.16 notes:
1. Fructose 2,6-bisphosphate is a potent *activator* of PFK1 and a potent *inhibitor* of FBP1, ensuring that glycolysis and gluconeogenesis are *not* on at the same time in the same cell.
2. PFK2 and FBP2 are in one bifunctional enzyme complex. When one is on, the other one is off.
3. FBP2 is the enzyme that is *hormonally* controlled. Glucagon (i.e., low blood sugar) activates FBP2; thus, PFK2 is off. Insulin (i.e., high blood sugar) inhibits FBP2; thus, PFK2 is now on.
4. *Hint*: When the **kinases are on** (PFK1 and PFK2), **glycolysis** is active in the cell. When the *bisphosphatases are on* (FBP1 and FBP2), *gluconeogenesis* is active in the cell.

and adenosine monophosphate (AMP). Allosteric inhibitors of PFK1 include ATP and citrate. Citrate is a TCA cycle intermediate but is transported into the cytosol for fatty acid synthesis (as will be covered in chapter 7). Citrate, then, serves to coordinate certain aspects of carbohydrate and lipid metabolism.

The energy state of the cell controls PFK1 and the flux of glucose through glycolysis. Since a purpose of the catabolism of glucose is to produce ATP, the major chemical form of energy for the cell, high cellular concentrations of ATP inhibit PFK1 as ATP binds to the allosteric sites on the enzyme. Glycolysis is inhibited, as one would expect. As ATP is consumed by cellular reactions, the concentrations of ADP and AMP rise in the cell. AMP serves as an activator of PFK1 and increases the flux of glucose through glycolysis to produce more ATP. In general, high levels of ATP in a cell result in the inhibition of catabolic processes, as a major purpose of catabolic processes is the generation of reducing power, leading to ATP production by the electron transport chain.

Pyruvate kinase, which catalyzes reaction 10 of glycolysis, is the third regulated step of glycolysis. Pyruvate kinase catalyzes the formation of ATP and pyruvate from phosphoenolpyruvate and ADP. Fructose 1,6-bisphosphate is an allosteric activator of pyruvate kinase. This activation by an intermediate formed earlier in the pathway is termed feed-forward activation. As this is the glycolytic step that results in the net yield of ATP, it is important that the pyruvate kinase is already active and ready to convert phosphoenolpyruvate to pyruvate to form ATP. High cellular concentrations of ATP also inhibit pyruvate kinase.

Additionally there are two isozymes of pyruvate kinase, the liver isozyme (L form) and the muscle isozyme (M form). The liver isozyme of pyruvate kinase is regulated by reversible phosphorylation of the enzyme, which is controlled by hormones such as insulin and glucagon. Insulin is the hormone that indicates the blood glucose level is high and stimulates the tissues that are targeted by insulin to take up glucose, as necessary, for their needs. Glucagon indicates the blood glucose level is low. The ratio of insulin to glucagon in the blood is important, more so than the individual concentrations of each hormone.

The liver is a target tissue of both insulin and glucagon. The liver must remove excess glucose from the blood, and it (along with the kidneys) is responsible for maintaining the blood glucose levels. The liver and kidneys are the only organs that have the enzyme glucose 6-phosphatase that allows glucose to be released back into the blood. Glucagon would lead to stimulation of gluconeogenesis (and inhibit glycolysis) in the liver, while insulin stimulates glycolysis (and inhibits gluconeogenesis).

Now consider when pyruvate kinase, a glycolytic enzyme, should be activated or inhibited. When gluconeogenesis occurs in a liver cell, pyruvate kinase needs to be off, or inhibited. Pyruvate kinase needs to be on, or active, during glycolysis. *Insulin* ultimately leads to the *dephosphorylation* of the L isozyme of pyruvate kinase to *activate* it because insulin activates the pathway of glycolysis because glucose is readily available. *Glucagon* ultimately leads to the *phosphorylation* of the L isozyme of pyruvate kinase to *inhibit* it because glucagon leads to the activation of gluconeogenesis because the liver must release glucose into the blood as blood glucose levels are dropping. In summary, the dephosphorylated form of pyruvate kinase is active, while the phosphorylated form of pyruvate kinase is inactive.

The next basic concept of regulation that will be covered is the reciprocal control of two enzymes, simultaneously, by one effector molecule. In this case, phosphofructokinase 1 (PFK1) of glycolysis and fructose 1,6-bisphosphatase (FBP1) of gluconeogenesis both utilize fructose 6-phosphate and fructose 1,6-bisphosphate (see **figure 2.16**, the reciprocal reactions drawn vertically on the left side of the figure). In *glycolysis*, PFK1 catalyzes the reaction of fructose 6-phosphate + ATP to from fructose 1,6-bisphospate + ADP. In *gluconeogenesis*, FBP1 catalyzes the dephosphorylation reaction of fructose 1,6-bisphosphate to form fructose 6-phosphate + P_i. Glycolysis and gluconeogenesis are reciprocal pathways, and therefore are not on in the same cell at the same time, as previously discussed.

How does the cell regulate which direction the reaction needs to go? Fructose 2,6-bisphosphate is a single effector molecule that controls both enzymes, PFK1 and FBP1—because they cannot both be on at the same time in the same cell. Fructose 2,6-bisphosphate, as its name implies, indicates it has phosphate groups on the hydroxy groups of carbons 2 and 6 of fructose (not carbons 1 and 6, as in fructose 1,6-bisphosphate). The structure of the effector molecule, fructose 2,6-bisphosphate, is shown in **figure 2.16** at the upper right of the figure. The production of this effector molecule is used solely for regulating PFK1 and FBP1. Fructose 2,6-bisphosphate is not an intermediate for other reactions of glycolysis or gluconeogenesis. Both PFK1 and FBP1 have allosteric binding sites for the molecule fructose 2,6-bisphosphate (i.e., not the catalytic sites of either enzyme). The purpose of fructose 2,6-bisphosphate is to bind to both of these enzymes to turn one of them on and the other off. To understand which one of them is on in the presence of fructose 2,6-bisphosphate, one needs to understand how this effector molecule is made and when the cell the is able to make it.

The synthesis and breakdown of fructose 2,6-bisphosphate is shown in **figure 2.16**, as the two reciprocal horizontal reactions at the top of the figure from left to right (and vice versa). Fructose 2,6-bisphosphate is made from fructose 6-phosphate. The phosphate group that is attached to the hydroxy group of carbon 2 comes from ATP. Enzymes that phosphorylate molecules using ATP are called *kinases* and are named after the acceptor of the phosphate—in this case fructose 6-phosphate. Hence, this enzyme is also call phosphofructokinase, but now we call it *phosphofructokinase 2* (PFK2). So now we have PFK1 and PFK2.

This effector molecule must also be removed from the cell, when necessary, to reverse which pathway is active or inhibited. The removal of fructose 2,6-bisphosphate will reverse which enzyme (i.e., PFK1 and FBP1) is active and which enzyme is inhibited. The formation of fructose 2,6-bisphosphate by PFK2 is also an irreversible reaction (just like the PFK1 catalyzed reaction). Therefore, a different enzyme is needed to remove the phosphate group from fructose 2,6-bisphosphate to re-form fructose 6-phosphate and inorganic phosphate (P_i). The enzyme that catalyzes the reverse reaction is a *phosphatase* named for the molecule that loses the phosphate group—*fructose 2,6-bisphosphatase* (FBP2). The enzyme names make sense. For PFK<u>1</u> and FBP<u>1</u> the enzymes are either adding a phosphate group or removing a phosphate group from <u>carbon 1</u> of a fructose derivative, respectively. For PFK<u>2</u> and FBP<u>2</u> the enzymes are either adding a phosphate group or removing a phosphate group from <u>carbon 2</u> of a fructose derivative, respectively. PFK2 and FBP2 actually exist together as an enzyme complex, in which one of them is hormonally controlled. FBP2 is hormonally controlled by reversible phosphorylation.

Now consider under what conditions should PFK1 and FBP1 be on or off. Fructose 2,6-bisphosphate is an effector molecule made from fructose 6-phosphate by a kinase that requires ATP. This effector molecule can only be made when there is plenty of glucose available because some of the fructose 6-phosphate will be diverted to make it. Therefore, the effector molecule fructose 2,6-bisphosphate should turn on PFK1 because the cell can do glycolysis when plenty of glucose is available. Fructose 2,6-bisphosphate binds to an allosteric site on PFK1 to activate it, and it binds to an allosteric site on FBP1 to inhibit it (shown in **figure 2.16** by the dotted lines with the (+) and (–) signs indicating activation or inhibition of the respective enzymes). Insulin is the hormone that indicates blood glucose levels are high and glucose is plentiful for tissues to take it up. Therefore *insulin* ultimately leads to the *dephosphorylation* of FBP2 to *inhibit* it. The inhibition of FBP2 results in the *activation of PFK2* to make the effector molecule fructose 2,6-bisphosphate because plenty of glucose is available to the cell.

When the cell needs to carry out de novo synthesis of glucose via gluconeogenesis, then all molecules that can become a glucose molecule need to be converted to glucose. Under these conditions, the effector fructose 2,6-bisphosphate molecules need to be converted back to fructose 6-phosphate molecules via FBP2. The fructose 6-phosphate will then continue through gluconeogenesis back to form glucose molecules. As fructose 2,6-bisphosphate is removed, PFK1 of glycolysis is inhibited and FBP1 of gluconeogenesis is activated. Glucagon is the hormone that signals low blood glucose levels, and the liver responds by releasing glucose into the blood. *Glucagon*, therefore, *phosphorylates* FBP2 to *activate* it, and *PFK2* will be *inhibited* under these conditions.

This complex reciprocal regulation by fructose 2,6-bisphosphate can be simplified into two rules:

1. When the kinases are on (meaning PFK1 and PFK2), the cell is doing glycolysis, and FBP1 and FBP2 are off under these conditions.
2. When the bisphosphatases are on (meaning FBP1 and FBP2), the cell is doing gluconeogenesis and the kinases, PFK1 and PFK2, are off.

The key point is that when the cell has two enzymes in the same compartment using the same intermediates, the regulation of both enzymes is important to directing the flux of intermediates into the necessary pathway.

METABOLIC FATES OF THE PRODUCTS OF GLYCOLYSIS

What happens to the products made in glycolysis, ATP, NADH, and pyruvate? ATP is the primary form of chemical energy for cells. The hydrolysis of ATP is used to drive many endergonic reactions and processes and is used in synthetic (anabolic) pathways, which generally require energy input to create larger molecules. One of the main purposes of glycolysis is that this catabolic pathway yields ATP directly, without the need for the electron transport chain.

Glycolysis also produces NADH and pyruvate. The fate of both of these molecules will depend on the availability of molecular oxygen (O_2). Molecular oxygen, though, is not used in any of the reactions of glycolysis. Why, then, would oxygen availability control the fate of products of glycolysis? Molecular oxygen is the final electron acceptor of the electron transport chain, located in the inner mitochondrial membrane. The reducing power of NADH and $FADH_2$, produced by catabolic pathways, are the electron donors to the electron transport chain. The movement of these electrons through the complexes of the electron transport chain, ending up on O_2 to reduce it to water, produces a proton gradient. The proton gradient is coupled to the phosphorylation of ADP by P_i to form ATP. The electron transport chain is the major pathway for the production of ATP from nutrient catabolism and requires reducing power from catabolic pathways and oxygen.

Recall that the pathway of glycolysis is in the cytosol. The NADH produced by the glyceraldehyde 3-phosphate dehydrogenase reaction is in the cytosol. The pools of cytosolic NAD^+/NADH and mitochondrial NAD^+/NADH are kept separate. The entire molecule of NAD^+ or NADH is not allowed to cross the inner mitochondrial membrane. Think of it in the following terms: If the cell sends the whole NADH molecule into the mitochondria to give its electrons to the electron transport chain, then the NAD^+ formed in the matrix will not come back out to the cytosol, because there are many more reactions in the mitochondria that use NAD^+ and NADH. However, the entire NADH molecule does not need to go the electron transport chain, just the electrons that it is carrying. The electrons on the cytosolic NADH formed in glycolysis will be transported to the electron transport chain using "shuttles." There are two shuttles, which will be discussed, the malate-aspartate shuttle and the glycerol phosphate shuttle. The shuttles only work under *aerobic* conditions, because that is when the electron transport chain is working and can use the electrons from the cytosolic NADH molecule. Under aerobic conditions, the shuttles are how cytosolic NAD^+ is regenerated for the glyceraldehyde 3-phosphate dehydrogenase step to continue.

Under aerobic conditions, pyruvate can also be transported into the mitochondrial matrix to continue the catabolism of pyruvate via the pyruvate dehydrogenase (PDH) complex and the tricarboxylic acid (TCA) cycle. The continued catabolism of pyruvate in the matrix will yield CO_2, reducing power (NADH and $FADH_2$), and ultimately more ATP via the electron transport chain.

Now consider what happens under anaerobic conditions, when there is not enough molecular oxygen to allow the electron transport chain to continue working. If the electron transport chain is inhibited, then the PDH complex and the TCA cycle in the mitochondria are also inhibited because the reducing power they produce cannot go to the electron transport chain. If the cell still needs to generate ATP under these conditions, glycolysis becomes the only catabolic process producing a net yield of ATP directly (via substrate-level phosphorylation reactions). The ten reactions of glycolysis still produce NADH (at the glyceraldehyde 3-phosphate dehydrogenase step) and pyruvate (as the product of the pyruvate kinase step). However, the electron transport chain is not operating under anaerobic conditions. As a result, the electrons from the cytosolic NADH cannot be shuttled to the electron transport chain, and pyruvate cannot be further catabolized by the PDH complex and the TCA cycle.

There is still the problem of regenerating cytosolic NAD^+. If glycolysis becomes the cell's only catabolic pathway generating ATP, a pool of cytosolic NAD^+ for the glyceraldehyde 3-phosphate dehydrogenase step still needs to be maintained. This is the importance of the location of glycolysis in the cytosol, as opposed to the mitochondrial matrix. If glycolysis was in the matrix, it would depend on the electron transport chain to regenerate its supply of NAD^+, as most other catabolic processes do.

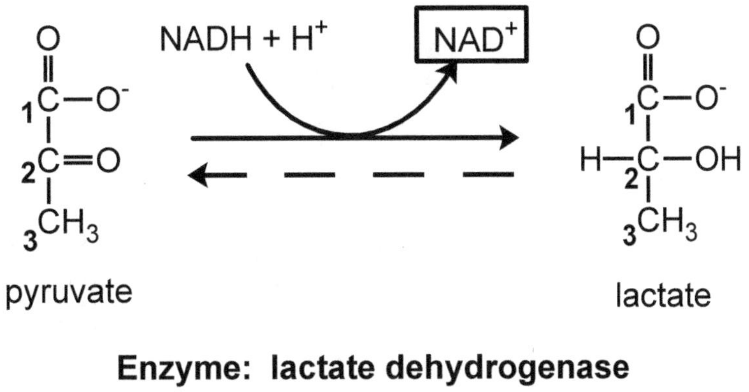

Enzyme: lactate dehydrogenase

Figure 2.17: Reaction 11: Formation of lactate from pyruvate (anaerobic conditions only, anaerobic glycolysis).

To regenerate cytosolic NAD$^+$ under anaerobic conditions, pyruvate stays in the cytosol, since it also cannot be transported into the mitochondria. Recall that pyruvate has a keto group on carbon 2. Under anaerobic conditions, glycolysis carries out an eleventh reaction (shown in **figure 2.17**), in which the keto group of pyruvate is reduced to form a secondary alcohol, producing the molecule lactate. This is a reduction reaction. If a molecule is reduced, a molecule must be oxidized. Normally, to go up the oxidation states flowchart (**figure 1.7**), the cell should use NADPH + H$^+$ and oxidize it to NADP$^+$. However, the sole purpose of this reaction is to regenerate cytosolic NAD$^+$ for glycolysis to continue. This reaction does, indeed, use *NADH + H$^+$* and oxidizes it to *NAD$^+$*. Enzymes that carry out redox reactions that use NAD$^+$/NADH or FAD/FADH$_2$ are called *dehydrogenases*. Dehydrogenases are named for the *more reduced* molecule, which is lactate. Thus, the enzyme name is *lactate dehydrogenase*. Now cytosolic NAD$^+$ is regenerated for the glyceraldehyde 3-phosphate dehydrogenase reaction, allowing glycolysis to continue to make ATP under anaerobic conditions when the electron transport chain is not working.

The lactate dehydrogenase reaction is a reversible reaction. The liver, however, is the only organ that can form pyruvate from lactate. All other tissues that generate lactate, especially muscle and red blood cells, will send the lactate back to the liver. Recall, as discussed previously, mature red blood cells do not have mitochondria and can only generate ATP by anaerobic glycolysis. The liver uses the lactate as a source of pyruvate for the pathway of gluconeogenesis to synthesize glucose, which is released into the blood and sent as an energy source for muscle and red blood cells. This process is called the *Cori cycle*.

SHUTTLE SYSTEMS FOR ELECTRON TRANSPORT

Under aerobic conditions, glycolysis has ten reactions and uses one of the shuttles to regenerate its cytosolic pool of NAD$^+$; while under anaerobic conditions, glycolysis consists of eleven reactions. The *malate-aspartate shuttle* and the *glycerol phosphate shuttle* are the two shuttles used by various tissues of the body.

What?
The malate-aspartate shuttle and the glycerol phosphate shuttle, under aerobic conditions, shuttle the *two electrons* via the hydride ion of cytosolic NADH into the mitochondria and generate either a mitochondrial NADH (for the malate-aspartate shuttle) or a mitochondrial $FADH_2$ (for the glycerol phosphate shuttle).

Why?
These shuttles have now regenerated cytosolic NAD^+ for the glyceraldehyde 3-phosphate dehydrogenase reaction so glycolysis can continue. The resulting mitochondrial NADH or $FADH_2$ now deliver these two electrons to the electron transport chain, which can be used to make ATP. These shuttles only transport the *two electrons* via hydrogen carriers from cytosolic NADH, not the whole molecule of NADH.

Where?
In the cell, these shuttles involve enzymatic reactions in both the cytosol and the mitochondria. The malate-aspartate shuttle is used by all tissues, and the heart and liver use it exclusively as it is the more energy-efficient shuttle. The glycerol phosphate shuttle is used by most tissues as a secondary mechanism for transporting electrons to the mitochondria.

How?
The malate-aspartate shuttle is shown in **figure 2.18**. As the purpose of the shuttle is to move electrons, the shuttle involves oxidation-reduction reactions. In **figure 2.18** begin with the molecules of NADH and oxaloacetate in the cytosol (lower left of the figure). Cytosolic malate dehydrogenase reduces oxaloacetate to malate with the simultaneous oxidation of NADH + H^+ back to NAD^+. The first goal of the shuttle has now been met—the regeneration of cytosolic NAD^+ for the glyceraldehyde 3-phosphate dehydrogenase reaction. Malate crosses into the mitochondrial matrix, where mitochondrial malate dehydrogenase oxidizes the malate to oxaloacetate and forms a mitochondrial molecule of NADH. (Note that this enzymatic reaction will be covered in more detail in the TCA cycle discussed in chapter 3.) This mitochondrial NADH can deliver its two electrons to the electron transport chain for the ultimate synthesis of ATP, the second goal of the shuttle.

There must be a pool of cytosolic oxaloacetate for this shuttle to operate, but oxaloacetate cannot cross the inner mitochondrial membrane. Therefore, a series of transamination reactions occur to regenerate the cytosolic pool of oxaloacetate. The relationship between α-keto acids and α-amino acids and their interconversion by transamination reactions was covered in chapter 1 (see **figure 1.5**). In the mitochondria, the α-amino group of glutamate is removed by the mitochondrial aspartate aminotransferase (AST), forming α-ketoglutarate. The mitochondrial AST then transfers the amino group to oxaloacetate to form the amino acid aspartate. The aspartate is transported to the cytosol, where a cytosolic AST reverses the transamination reaction to generate cytosolic oxaloacetate.

The malate-aspartate shuttle is readily *reversible* (unlike the glycerol phosphate shuttle). Thus, cytosolic NADH electrons can only be transferred to the mitochondria when the $NADH/NAD^+$ ratio is higher in the cytosol than in the mitochondria. Gluconeogenesis (covered in chapter 5), in fact, depends on this shuttle running in reverse to move the mitochondrial oxaloacetate formed by pyruvate carboxylase to the cytosol by reducing it to malate.

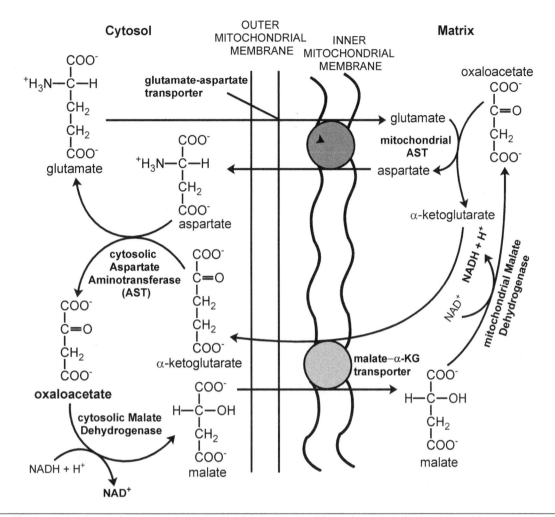

Figure 2.18: The malate-aspartate shuttle. (Note: Start with oxaloacetate in the cytosol.)

The glycerol phosphate shuttle is shown in **figure 2.19**. In this shuttle, cytosolic dihydroxyacetone phosphate is needed. Recall that dihydroxyacetone phosphate is the glycolytic intermediate formed by reaction 4 of glycolysis when fructose 1,6-bisphosphate is cut in half. In the first reaction of this shuttle, cytosolic *glycerol phosphate dehydrogenase* (a.k.a. *glycerophosphate dehydrogenase*) reduces the keto group of dihydroxyacetone to form glycerol phosphate (whose name indicates its structure) with the simultaneous oxidation of cytosolic NADH + H$^+$ back to NAD$^+$. Note that the enzyme is correctly named as a *dehydrogenase*, named for the *more reduced* molecule. Now the first goal of the shuttle, the regeneration of cytosolic NAD$^+$ for the glyceraldehyde 3-phosphate dehydrogenase reaction of glycolysis, has been completed.

Glycerol phosphate now moves into the intermembrane space of the mitochondria. Note that glycerol phosphate does not cross the mitochondrial inner membrane to go into the matrix. The mitochondrial glycerol phosphate dehydrogenase is embedded in the outer face of the mitochondrial inner membrane. It oxidizes glycerol phosphate back to dihydroxyacetone phosphate, which goes back to the cytosol. However, the

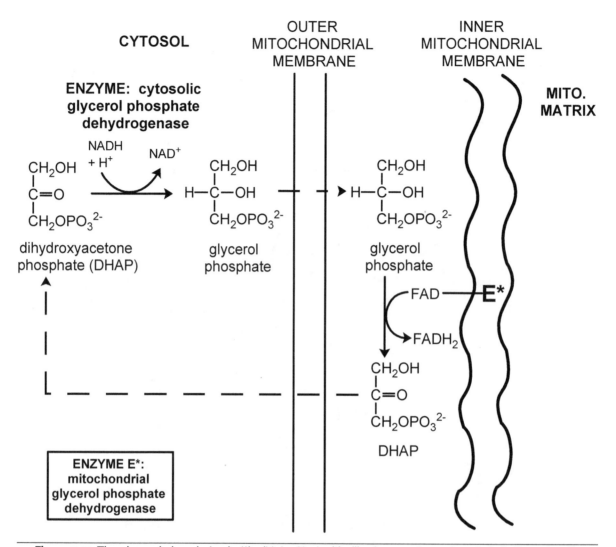

Figure 2.19: The glycerol phosphate shuttle. (Note: Start with dihydroxyacetone phosphate in the cytosol.)

mitochondrial glycerol phosphate dehydrogenase uses *FAD* as its coenzyme and reduces it to *FADH$_2$*. The use of FAD for this reaction is an exception to the general rule for when to use FAD, but it serves a key purpose for this shuttle. The electrons of this FADH$_2$ are passed directly to coenzyme Q, an important intermediate of the electron transport chain, and thus are used for ATP synthesis—the second goal of the shuttle.

For the glycerol phosphate shuttle, the net conversion of cytosolic NADH to mitochondrial FADH$_2$ seems to waste about one ATP. The advantage of this shuttle, though, is that it allows the electrons from cytosolic NADH to be transported to the mitochondria against an NADH concentration gradient. The transport of the electrons from cytosolic NADH against the concentration gradient is especially important in actively metabolizing muscle tissue. In actively metabolizing tissues, the catabolic processes occurring in the mitochondria are producing much more NADH than is being produced by glycolysis in the cytosol.

SUMMARY SHEET OF GLYCOLYSIS

To complete the summary sheet, draw the indicated structures.

1. **Who?** Humans, and most living organisms, including bacteria, plants, and other animals.
2. **What?** [Definition] Conversion of glucose (a six-carbon unit) to two pyruvates (three-carbon units) under aerobic conditions, or conversion to two lactates (three-carbon units) under anaerobic conditions.
3. **Why?** [Purpose] To produce energy (ATP) and reducing power (NADH).
4. **Where?** Cytosol of the cell.
5. **How?**
   a. Start with glucose (*draw the structure*)

   b. End with pyruvate or lactate (*draw both structures*)

   c. Key intermediates of the pathway: glyceraldehyde 3-phosphate and dihydroxyacetone phosphate (*draw both structures*)

   d. Committed step (*draw underlined structures*)
      (1) Fructose 6-phosphate + ATP → fructose 1,6-bisphosphate + ADP

      (2) Catalyzed by phosphofructokinase (PFK; PFK1)
      (3) Step is irreversible and highly regulated
   e. Four kinase reactions
      (1) Two use ATP (hexokinase/glucokinase; phosphofructokinase)
      (2) Two produce ATP (phosphoglycerate kinase; pyruvate kinase)
   f. Two possible oxidation-reduction reactions
      (1) One produces NADH for both aerobic and anaerobic glycolysis (glyceraldehyde 3-phosphate dehydrogenase)
      (2) One uses NADH, only during anaerobic glycolysis (lactate dehydrogenase)
   g. Shuttles transport electrons of cytosolic NADH to electron transport chain for aerobic glycolysis
6. **When?**
   a. Three irreversible reactions of glycolysis, which are all regulated steps

b. Reaction 1: Hexokinase (note: isozyme glucokinase, regulated differently)
      (1) Reaction 1: glucose + ATP → glucose 6-phosphate + ADP
      (2) Hexokinase: inhibited by glucose 6-phosphate (glucokinase is not inhibited by glucose 6-phosphate)
   c. Reaction 3: Phosphofructokinase 1
      (1) Reaction 3: fructose 6-phosphate + ATP → fructose 1,6-bisphosphate + ADP
      (2) Activated by: fructose 2,6-bisphosphate; AMP; insulin (via FBP2, signals plenty of glucose available)
      (3) Inhibited by: ATP and citrate; glucagon (via FBP2, signals low blood glucose)
   d. Reaction 10: Pyruvate kinase
      (1) Reaction 10: phosphoenolpyruvate + ADP → pyruvate + ATP
      (2) Activated by: fructose 1,6-bisphosphate; insulin
      (3) Inhibited by: ATP; glucagon
7. Review of entire pathway.
   a. First three steps phosphorylate carbons 1 and 6 as primary alcohols using ATP
      (1) Glucose + ATP → glucose 6-phosphate + ADP (carbon 6 is a primary alcohol); **hexokinase or glucokinase**
      (2) Isomerize glucose 6-phosphate to fructose 6-phosphate (now carbon 1 is a primary alcohol); **phosphoglucose isomerase**
      (3) Fructose 6-phosphate + ATP → fructose 1,6-bisphosphate + ADP; **phosphofructokinase**
   b. Now split fructose 1,6-bisphosphate into two three-carbon phosphorylated intermediates and isomerize the ketone to the aldehyde
      (1) Fructose 1,6-bisphosphate → dihydroxyacetone phosphate + glyceraldehyde 3-phosphate; **aldolase**
      (2) Dihydroxyacetone phosphate ↔ glyceraldehyde 3-phosphate (aldehyde group can be oxidized); **triose phosphate isomerase**
   c. Oxidize the aldehyde group of glyceraldehyde 3-phosphate, reduce NAD^+, and phosphorylate the acid group using inorganic phosphate
      (1) Glyceraldehyde 3-phosphate + NAD^+ + P_i → 1,3-bisphosphoglycerate + NADH + H^+; **glyceraldehyde 3-phosphate dehydrogenase**
   d. Use the phosphate group on carbon 1 to make first ATP (because this phosphate is a good leaving group and has enough energy to make a phosphoanhydride bond)
      (1) 1,3-bisphosphoglycerate + ADP → 3-phosphoglycerate + ATP; **phosphoglycerate kinase**
   e. Reactions 8, 9, and 10 make the phosphate group on carbon 3 a good leaving group to make ATP, and position a keto group next to the carboxylic acid group to make it a good leaving group as CO_2 (by the PDH complex)
      (1) 3-phosphoglycerate → 2-phosphoglycerate; **phosphoglycerate mutase**
      (2) 2-phosphoglycerate → phosphoenolpyruvate + H_2O; **enolase**
      (3) phosphoenolpyruvate + ADP → pyruvate + ATP; **pyruvate kinase**
   f. Regeneration of cytosolic NAD^+ for use by glyceraldehyde 3-phosphate dehydrogenase
      (1) Aerobic conditions: use shuttles to move the electrons from cytosolic NADH to the electron transport chain, regenerating cytosolic NAD^+
      (2) Anaerobic conditions: pyruvate + NADH + H^+ → lactate + NAD^+; **lactate dehydrogenase**

CHAPTER 3

MITOCHONDRION OVERVIEW, THE PYRUVATE DEHYDROGENASE COMPLEX, AND THE TRICARBOXYLIC ACID CYCLE

OBJECTIVES

1. Explain the purpose of the PDH complex and why it is a key regulatory point of metabolism.
    a. Identify the aerobic and anaerobic fates of pyruvate.
2. Identify where the PDH complex is located in the cell.
3. Explain the overall reaction steps (as fits the oxidation states pattern of metabolism) of the PDH complex and how certain vitamin deficiencies would affect this complex.
    a. Identify the two main steps of the PDH complex as applications of the oxidation states basic principles.
    b. Identify the three catalytic components of the PDH complex, each of their cofactors, and their vitamin precursors.
4. Explain when the PDH complex is activated or inhibited.
    a. Explain the regulation of the PDH complex in terms of both of its regulatory components (the inhibitors and activators of the complex) and how it is hormonally controlled.
5. Define the process of the TCA cycle.
6. Explain the purpose of the TCA cycle.
    a. Describe the role of the TCA cycle in energy production.
7. Identify where the TCA cycle takes place in the cell.
8. Explain how the TCA cycle is carried out in the cell and how vitamin deficiencies would affect this pathway.
    a. Name and recognize the structures of the starting products, key intermediates, and end products of the TCA cycle.
    b. Name the eight enzymes involved in the TCA cycle and describe the reactions they catalyze.
    c. Identify the four reactions of the TCA cycle in which reducing equivalents are produced.
    d. Identify the products per turn of the TCA cycle and be able to differentiate why glucose fuels two turns but alanine can only fuel one turn of the TCA cycle.
9. Explain when the TCA cycle occurs.
    a. Identify the regulatory enzymes of the TCA cycle, how intermediates can be used for other synthetic processes, and how anaplerotic reactions can replenish TCA cycle intermediates.
10. Explain how a thiamine deficiency will affect the PDH complex and the TCA cycle and why it would cause elevated levels of pyruvate and lactate in the blood.

MITOCHONDRIA: STRUCTURE AND FUNCTION

To get the maximum energy yield of ATP from glucose (thirty to thirty-two ATP), pyruvate must be transported into the mitochondria, a cellular organelle. There the pyruvate will be completely oxidized to CO_2 and will make more reducing power (NADH and $FADH_2$) in the process. The reducing power will be used by the electron transport chain to produce a proton gradient. The proton gradient (i.e., electrical energy) will then be converted to chemical energy in the form of ATP.

A mitochondrion is about one micron in size, which is about the size of an *Escherichia coli* bacterium (see **figure 3.1**). A mitochondrion has an inner and outer membrane, an intermembrane space, and a matrix. The outer membrane of a mitochondrion contains pores, which are voltage-dependent anion channels. These pores are permeable to many small molecules, including ATP, ADP, and many metabolites. The intermembrane space, between the two membranes of the mitochondrion, is where the protons are pumped by the electron transport chain.

The inner membrane is highly folded and forms a series of internal ridges referred to as cristae. These cristae can extend across the mitochondrial matrix to essentially create separate compartments within the matrix. All five complexes of the electron transport chain are also located in the inner membrane.

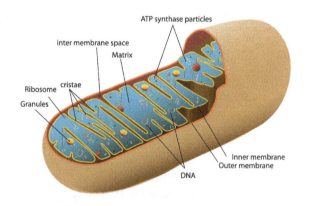

Figure 3.1: Basic structure of a mitochondrion.

The inner membrane is impermeable to nearly all ions and polar molecules. If the inner membrane will not let protons (the smallest atom) freely flow across the membrane, nothing else is going to go across either. Thus, anything that is allowed to cross the inner membrane is controlled and will have to have some sort of transporter located in the inner membrane (as shown in **figure 3.2**). **Figure 3.2** shows examples of the concept that anything that crosses the inner mitochondrial membrane needs a specific transporter. Think of the transporter as a door. If the membrane is going to open the door to let a molecule in, then it also lets a molecule out. Thus, the organelle maximizes the use of any force required to open that door.

Thus, a pair of molecules typically pass through a particular door. When one of the transporters (i.e., doors) opens, a molecule will exit the mitochondrion, and a molecule will come into the mitochondrial matrix. Malate is a molecule that is allowed to cross the inner membrane, and it has a several transporters that let it in or out of the matrix. Aspartate, glutamate, and α-ketoglutarate are also allowed to cross and are key transporters of the malate-aspartate shuttle involved in regenerating cytosolic NAD^+ for glycolysis (discussed in chapter 2). Notice, though, that oxaloacetate, NADH, and NAD^+ are not listed in the diagram. These molecules are not allowed to cross the inner membrane. There are numerous ion transporters, as well. The

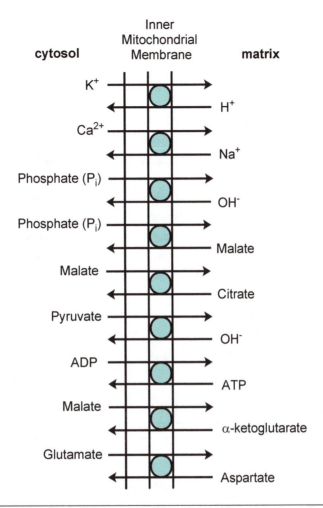

Figure 3.2: Inner membrane transporters. Various inner membrane transporters control the flow of essential molecules into and out of the matrix. The primary goal of these inner membrane transporters is to establish or maintain the membrane potential.

primary goal of these transporters is to establish or maintain the membrane potential, as well as transport intermediates necessary for various metabolic pathways.

The matrix of the mitochondrion, the center region of the organelle, carries out many pathways. Mitochondria are primarily catabolic organelles, as the primary function of mitochondria is the production of ATP using the reducing power generated by the breakdown (catabolism) of many molecules. The pyruvate dehydrogenase (PDH) complex and the tricarboxylic acid (TCA) cycle are located in mitochondria. The TCA cycle is the common destination of intermediates from the catabolism of carbohydrates, proteins, lipids, and nucleotides. Fatty acid oxidation (β-oxidation) is another breakdown pathway in the matrix. There are, however, some synthetic pathways in the matrix. These include ketone body synthesis, heme synthesis, and steroid synthesis. Some pathways actually bridge the mitochondrion. Part of the pathway will be in the mitochondrion, and part will take place in the cytosol. For example, gluconeogenesis is a synthetic pathway. The initial reactions of gluconeogenesis are actually in the mitochondrial matrix, while the remainder of the pathway is in the cytosol. As a result of the numerous pathways, the matrix has a very high concentration of proteins. The mitochondrion also has its own genome (located in the matrix). There are various diseases that occur in humans due to mutations in the mitochondrial genome.

Mitochondria use greater than 90 percent of the oxygen we breathe, as oxygen is the terminal electron acceptor of the electron transport chain. Oxygen is needed by all of the mitochondria because the mitochondria produce greater than 90 percent of all the ATP, which is a major function of the mitochondria. All of the catabolic pathways in the mitochondria (i.e., the TCA cycle, β-oxidation, and others) send all of their NADH and $FADH_2$ to the electron transport chain to regenerate the oxidized form of those molecules (NAD^+ and FAD). If the electron transport chain is not working, the cell does not run the PDH complex, the TCA cycle or β-oxidation, because there is nowhere for all that reducing power to go. Glycolysis is the single *catabolic* pathway that will produce ATP, without relying on the electron transport chain to regenerate its cytosolic NAD^+. Again, it is very important that glycolysis is *not* in the mitochondria. When cells lack oxygen, glycolysis produces ATP for the cell by anaerobic glycolysis. As previously mentioned, mature red blood cells lack mitochondria and can only produce ATP via anaerobic glycolysis. Since the purpose of red blood cells is to transport oxygen to other cells and tissues, it is important that they do not use what they are transporting.

THE PYRUVATE DEHYDROGENASE COMPLEX: THE BRIDGING STEP BETWEEN GLYCOLYSIS AND THE TCA CYCLE

The pyruvate dehydrogenase (PDH) complex is the *irreversible* bridging step between glycolysis and the TCA cycle, which take place in the cytosol and the mitochondria respectively. Recall that α-keto acids are stable, such that the carboxylic acid group does not readily leave the molecule. Therefore, decarboxylating an α-keto acid is mechanistically hard or complex. Presented here is a *very* simplified reaction sequence for this complex. It is not the actual mechanism. The intermediate drawn is not a real intermediate of this complex. The point is to draw the reactions so that they line up with what one has learned in the basic recipe of metabolism (**figure 1.7** of chapter 1). The PDH complex is a multienzyme complex that converts pyruvate into acetyl CoA. Acetyl CoA is a structure one must know (recognize and be able to draw the two-carbon part, not necessarily the CoA part). As the PDH complex is irreversible, humans can convert glucose (and other sugars) to fatty acids, and ultimately triacylglycerols, but cannot convert fatty acids back to glucose. This will be discussed further in covering gluconeogenesis in chapter 5.

The PDH complex can be defined (*what*) as a multienzyme complex that converts pyruvate (a three-carbon molecule) to acetyl CoA (a two-carbon unit attached to coenzyme A) for entry into the TCA cycle under *aerobic* conditions. The net reaction for the PDH complex is shown in **figure 3.3**. The products of the PDH complex are acetyl CoA, NADH, and CO_2. The reason the reaction is irreversible is due to the negative $\Delta G°'$ equal to –8 kcal/mol for the net reaction of the complex. For a ΔG of a reaction that is equal to –4 kcal/mol, or is more negative than –4 kcal/mol, there is essentially nothing the cell can do to manipulate

$$1 \text{ pyruvate} + \text{CoA} + \text{NAD}^+ \longrightarrow 1 \text{ acetyl CoA} + CO_2 + \text{NADH} \qquad \Delta G^{0\prime} = -8 \text{ kcal/mol}$$

Figure 3.3: Net reaction of the PDH complex.

the ratio of products to reactants to make the reaction go in the other direction. Thus, it is an irreversible reaction. In biological systems, though, certain irreversible reactions must be bypassed in some manner (e.g., gluconeogenesis bypassing the three irreversible reactions of glycolysis). The PDH complex, however, is an irreversible reaction that is not bypassed by another set of reactions.

The purpose of the PDH complex (*why*) is to decarboxylate pyruvate to form CO_2. The formation of CO_2 allows for the excretion of the first carbons from the breakdown of glucose (or other monosaccharides like fructose and galactose) or from the breakdown of certain amino acids. The PDH complex also produces NADH, as shown in the net reaction in **figure 3.3**. This NADH is formed in the mitochondrial matrix, where the complex is located, so it can go directly to the electron transport chain for energy production. The electron transport chain will then regenerate mitochondrial NAD^+ for the pathways located in the matrix.

How does the PDH complex work? Again, think of this whole series of reactions as "hard" because it is decarboxylating an α-keto acid. The actual enzyme complex is large and contains five different enzymes, three catalytic and two regulatory. **Table 3.1** indicates the names of the three catalytic components of the complex, though they are often simply referred to as E_1, E_2, and E_3. In naming enzymes, anytime an enzyme (or enzyme complex) carries out multiple reactions, if one of the reactions is a redox reaction using NAD^+ or FAD, then the enzyme is called a *dehydrogenase*. Dehydrogenases are named for the *more reduced* molecule. However, the actual reduced molecule of the oxidation-reduction reaction of the complex is an intermediate attached lipoamide. Therefore, the enzyme name just uses the starting material of the complex, pyruvate. Hence, the name of this enzyme complex is the *pyruvate dehydrogenase complex*. In fact, the first catalytic enzyme component is called pyruvate dehydrogenase.

Each of the three catalytic enzymes has a prosthetic group associated with it (see **table 3.1**). A prosthetic group is a coenzyme that is permanently associated with an enzyme. While a simplified reaction scheme for the complex will be drawn, knowing the prosthetic groups of each of the catalytic subunits is important. These prosthetic groups (coenzymes) are derived from vitamins. Thus, certain vitamin deficiencies will

Table 3.1. The three PDH complex catalytic enzymes and their overall reactions

ENZYME	ABBREVIATION	PROSTHETIC GROUP	REACTION CATALYZED
Pyruvate dehydrogenase	E_1	TPP (thiamine pyrophosphate)	Oxidative decarboxylation of pyruvate
Dihydrolipoyl transacetylase	E_2	Lipoamide	Transfer of the acetyl group to CoA
Dihydrolipoyl dehydrogenase	E_3	FAD	Regeneration of the oxidized form of lipoamide

affect the function of this complex, especially as it requires a number of coenzymes to function, as indicated in the chart. The first one is thiamine pyrophosphate (TPP), which is derived from the vitamin thiamine (B_1). Lipoamide is not a vitamin because the body actually makes it. However, flavin adenine dinucleotide (FAD) is derived from the vitamin riboflavin (B_2). Coenzyme A is part of the product of the complex, acetyl CoA. Coenzyme A is derived from the vitamin pantothenic acid (B_5). The vitamin niacin (B_3) is needed to produce the coenzyme nicotinamide adenine dinucleotide (NAD^+), which will be reduced to produce $NADH + H^+$. A deficiency in thiamine, riboflavin, or niacin will affect the ability of the PDH complex to function, as well as the α-ketoglutarate dehydrogenase complex of the TCA cycle because its mechanism is the same as the PDH complex.

Table 3.1 indicates the overall reaction catalyzed by each component. Note that E_3 has nothing to do with converting pyruvate to acetyl CoA. Lipoamide of E_2 is reduced as the product of the E_1 reaction, and the purpose of E_3 is to re-oxidize the lipoamide. Thus, the complex catalyzes a series of oxidation-reduction reactions, as well as the decarboxylation that takes place at E_1. For the catalytic reaction of E_3, its prosthetic group FAD gets reduced to $FADH_2$. This $FADH_2$ cannot go to the electron transport chain because it is permanently attached to E_2. For the whole complex to function again, the $FADH_2$ of E_3 needs to be re-oxidized back to FAD. Thus this is where NAD^+ is reduced to $NADH + H^+$. This NADH is not attached to the complex and does carry electrons to the electron transport chain.

The advantage of having all of these enzymes together in a single complex enhances the rate of the sequential reactions and minimizes the side reactions that could occur. The organization of the enzymes into a complex also serves to stabilize against proteolysis, the breakdown of the enzyme itself. Coordinated regulation is also permitted by this organization of enzymes, as all of the reactions need to occur or none of them. Again, this is an irreversible step between glycolysis and the TCA cycle, so regulation of the complex is very important.

Figure 3.4 is the simplified reaction sequence carried out by the PDH complex. It is divided into a two-step reaction. Pyruvate has a carboxylic acid group, with a keto group in the alpha position. The first step is to decarboxylate pyruvate, to yield what the author terms an imaginary or fake intermediate, acetaldehyde. The aldehyde group would be formed if the intermediate were to be released. In actuality the intermediate is attached to the lipoamide coenzyme of E_2, so the aldehyde functionality is not really formed.

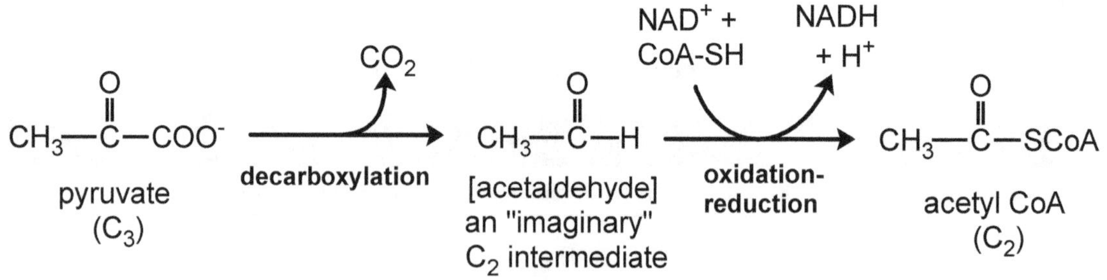

Figure 3.4: The PDH complex reaction as two main steps (as applies to the oxidation states flowchart). Note: The real mechanism is much more complex. The main point is that α-keto acids are able to be decarboxylated but require a complex mechanistic reaction sequence.

In this simplified reaction scheme, the first step is decarboxylation, and the second step is a redox reaction. This two-step reaction scheme will account for all of the products of the PDH complex. Following the oxidation states flowchart, after the decarboxylation forms the aldehyde, the complex oxidizes it. Oxidation of an aldehyde yields a carboxylic acid, but in this case the carboxylic acid gets attached to coenzyme A to form a thioester. A thioester function is equivalent in its oxidation state to a carboxylic acid. So attaching the acetaldehyde at the aldehyde carbon to CoA elevates it in oxidation state to the oxidation state of a carboxylic acid. Another way to look at it is if the CoA were to be clipped off by adding water across the thioester bond, the product would be acetate (a two-carbon fatty acid).

As noted in the discussion of the three catalytic enzymes, the real mechanism is much more complex. However, this simple two-step reaction sequence follows the rules of the oxidation states flowchart and accounts for all the products of the complex. Oxidation and reduction reactions are always coupled. According to the rules, NAD^+ is used when oxidizing an aldehyde to an acid, which accounts for the product $NADH + H^+$. (The real reaction series involves the oxidation and reduction of two other prosthetic groups, lipoamide and FAD.) It is the NADH formed, though, that is able to carry electrons to the electron transport chain. The PDH complex decarboxylates pyruvate, forming CO_2, acetyl CoA, and NADH. These are the products of this complex, and the starting and ending materials of pyruvate and acetyl CoA are two key structures one should know.

This simplified two-step reaction scheme will be shown again with the α-ketoglutarate dehydrogenase complex in the TCA cycle. The mechanism is the same because the complex is also decarboxylating an α-keto acid, exactly like the PDH complex. However, the α-ketoglutarate dehydrogenase complex does not have the same regulatory subunits on it. It does have the same prosthetic groups, the same enzyme mechanism, and the product is also attached to CoA.

Regulation of pathways is very complex. Control of metabolic pathways is often more dependent on the overall ratio of key regulators, rather than the absolute amounts of any given regulator. In this text, regulation of pathways and enzymes will be covered to emphasize several major concepts of regulation. The first is that coordinated regulation of pathways often includes the reversible phosphorylation of key enzymes. The PDH complex has two regulatory enzymes that control the phosphorylation state of E_1 (the pyruvate dehydrogenase component), since that is where the reaction series starts. The two regulatory enzymes of the PDH complex are the *pyruvate dehydrogenase kinase* and the *pyruvate dehydrogenase phosphatase*. Recall that kinases are enzymes that phosphorylate something and are named after what they phosphorylate. Thus, the pyruvate dehydrogenase kinase phosphorylates the pyruvate dehydrogenase component (E_1). Phosphorylation of E_1 turns *off* the complex. Removal of this phosphate group by the pyruvate dehydrogenase phosphatase would then turn on the PDH complex. Note the phosphatase is named after what it dephosphorylates.

Insulin and glucagon are two hormones that control the phosphorylation state of the PDH complex in their respective target tissues (as well as key enzymes in other metabolic pathways) to achieve coordinated control of these metabolic pathways. Insulin, as mentioned previously, is the hormone that indicates blood glucose is plenty, and cells can take up glucose as needed to carry out metabolic pathways. Insulin is considered an activator of synthetic pathways, since glucose is readily available as an energy source through its catabolism. Glucose may also be used in synthetic pathways and to replenish glycogen stores for certain cells/tissues. Glucagon, on the other hand, indicates low blood sugar and is responsible for targeting the liver (primarily), resulting in the release of glucose into the blood from the liver either from liver glycogen stores or from gluconeogenesis (de novo synthesis of glucose).

One should consider what enzymes should be on or off in the liver when insulin is the predominant hormone in the blood versus glucagon. Again, the liver is the organ responsible for sensing the nutrient needs of the body and supplying them. For the PDH complex, consider which hormone would active it, and which hormone would inactivate it? If one has just eaten a meal with abundant carbohydrates, the pancreas would secrete insulin indicating glucose is readily available. The liver needs to respond to the abundant glucose coming in from digestion of the meal to prevent the blood glucose levels from becoming too high. The liver is responsible for removing the excess nutrients and using them or storing them. As excess glucose is detected by the liver, the liver must remove the glucose from the blood and do something with it. Initially liver cells may need to use the glucose to replenish their own ATP pools by carrying out glycolysis, the PDH complex, the TCA cycle, and the electron transport chain. Additional glucose may be used by liver cells to replenish their storage of glucose as glycogen. Once the glycogen stores are replenished, further glucose will be converted to fatty acids (discussed in chapter 7) and stored as triacylglycerols. Insulin would activate the PDH complex under these conditions. Insulin would therefore activate the *pyruvate dehydrogenase phosphatase* to dephosphorylate E_1 to turn on the PDH complex.

When glucagon is the predominant hormone, the liver will begin by releasing glucose into the blood from its glycogen stores. As the glycogen stores are used, glucose will be made de novo from pyruvate via gluconeogenesis. If gluconeogenesis is activated to help replenish blood glucose levels, then the PDH complex needs to be off (as the pyruvate carboxylase enzyme of gluconeogenesis is on). Thus, glucagon activates the *pyruvate dehydrogenase kinase* to phosphorylate E_1 of the PDH complex to turn it off.

Another key regulatory concept one should consider is the ratio of the cellular concentration of ATP versus ADP or AMP. If the concentration of ATP is high in the cell, the cell is generally doing synthetic processes, not catabolic processes. The PDH complex would generally be inhibited by high ATP concentrations and activated by high AMP concentrations.

A third key regulatory concept involves two enzymes in the same cellular compartment that use the same substrate. In mitochondria there are two enzymes that utilize pyruvate as a substrate, the PDH complex and pyruvate carboxylase. These two enzymes are not on simultaneously (in general) in the mitochondria. Again the focus is on the liver, as a primary example. If the liver is doing gluconeogenesis utilizing pyruvate carboxylase (discussed in chapter 5), the PDH complex needs to be off. To achieve this control, acetyl CoA inhibits the PDH complex and activates pyruvate carboxylase. Thus, control of two enzymes is achieved with one molecule. One can attribute the inhibition by acetyl CoA as product inhibition of the PDH complex. High concentrations of acetyl CoA could result if the TCA cycle is slowing down because the cell has plenty of ATP. If the electron transport chain is off, the concentrations of NADH build up in the cell, leading to the inhibition of the TCA cycle and ultimately the PDH complex because the acetyl CoA is not entering the TCA cycle. Particularly in the liver, a high concentration of acetyl CoA generally comes from the β-oxidation of fatty acids (discussed in chapter 6), and the acetyl CoA will be used for ketone body synthesis (rather than go into the TCA cycle) under those conditions. The importance of acetyl CoA as a regulator of pyruvate carboxylase and the PDH complex will be discussed in more detail when covering the lipid pathways.

Again, the PDH complex is a bridging step and a key regulatory point for the fate of pyruvate. The conversion of pyruvate to acetyl CoA is irreversible, and in humans fatty acids cannot be converted back to sugars. Thus, the conversion of glucose ultimately to acetyl CoA (i.e., the PDH complex is on) typically commits acetyl CoA molecules to one of two paths. They may enter the TCA cycle to burn off the two carbons of the acetyl CoA as CO_2 and produce more reducing power in the form of NADH and $FADH_2$ for the electron transport chain for energy production. Or the acetyl CoA may be used for lipid synthesis (i.e., fatty acids for triacylglycerides or membrane lipids, or cholesterol). If the PDH complex is off, pyruvate formed from glycolysis will be used for lactate production (i.e., anaerobic glycolysis) or for amino acid synthesis, like alanine. If pyruvate is converted to oxaloacetate by pyruvate carboxylase in the mitochondria, then gluconeogenesis is occurring and the PDH complex is off under these conditions.

THE TRICARBOXYLIC ACID CYCLE

The tricarboxylic acid (TCA) cycle, also referred to as the *citric acid cycle* or the *Krebs cycle*, is a pathway that truly demonstrates all of the principles covered in the oxidation states chapter (chapter 1). This pathway will provide a concrete example of how to understand a pathway by applying the oxidation states flowchart (**figure 1.7**), as well as the enzyme naming rules, functional groups that cannot be further oxidized, and the removal of carboxylic acid groups from a molecule.

Since this pathway is a cycle, one should learn the products of the pathway "per cycle" or "per turn" of the cycle, as shown in the net reaction of **figure 3.5**. The catabolic pathways of various biological molecules funnel intermediates into the TCA cycle. Thus, one should know which reactions produce the products of the cycle and that the entry of acetyl CoA into the cycle yields all of the products. This will help in understanding the energy yields obtained from the catabolism of various nutrients. For example, a molecule of glucose will ultimately turn the TCA cycle two times (i.e., the catabolism of glucose ultimately yields two molecules of acetyl CoA), but a fatty acid will turn the TCA cycle many times (i.e., palmitate, a C_{16} fatty

$$1 \text{ acetyl CoA} + 3 \text{ NAD}^+ + \text{FAD} + \text{GDP} + P_i + 2 H_2O \longrightarrow 2 CO_2 + 3 \text{ NADH} + FADH_2 + \text{GTP} + 2 H^+ + \text{CoA}$$

Figure 3.5: The net reaction of the TCA cycle. These are the products "per turn" of the TCA cycle.

acid, will turn it eight times). The catabolism of a molecule of alanine, by transamination to a molecule of pyruvate that is then converted to acetyl CoA by the PDH complex, can only turn the TCA cycle one time.

For the net reaction of the TCA cycle (**figure 3.5**), the entry of acetyl CoA, a two-carbon unit, into the pathway is considered the start of the cycle. The products of the TCA cycle include two CO_2 molecules. Those exact two carbons of the acetyl CoA entering the cycle are not burned off in a single turn of the cycle. It usually takes a couple of turns to burn off those two specific carbons. The focus here, though, is not on the actual mechanisms—rather, the pattern. Thus, two carbons are coming into the cycle, and two carbons are burned off. Acetyl CoA, then, cannot yield a glucose molecule. The PDH complex is irreversible, and the entry of acetyl CoA into the TCA burns off two carbons prior to the formation of oxaloacetate. Also the formation of oxaloacetate for gluconeogenesis comes from pyruvate via pyruvate carboxylase, not acetyl CoA. Thus, the cell cannot achieve net synthesis of glucose from a molecule of acetyl CoA. The TCA cycle also yields four molecules of reducing power (three NADH and one $FADH_2$), which go to the electron transport chain for ATP synthesis. In addition, the TCA cycle yields one GTP (an energy equivalent) directly, some protons, and a molecule of coenzyme A.

To define the TCA cycle (*what*): The TCA cycle completes the oxidation of the acetyl unit (the two-carbon portion) of acetyl CoA to CO_2 and serves as the final common pathway for the oxidation of molecules (carbohydrates, fatty acids, amino acids, and nucleotides). The two molecules of pyruvate from glycolysis have been converted to two molecules of acetyl CoA along with the release of 2 CO_2. Now there are four carbons left from the original glucose molecule on these two acetyl CoA molecules. Each of these four carbons needs to be oxidized to the level of a carboxylic acid. They can then be clipped off as CO_2 to complete the catabolism of the six carbons of glucose. These two molecules of acetyl CoA formed from glucose will turn the TCA cycle two times. By the end of two turns of the TCA cycle, four CO_2 molecules will be formed, essentially completing the oxidation of glucose. (Note, as mentioned previously, that one turn of the TCA cycle does not clip off the same two carbons from the entering acetyl unit.) The catabolism of fatty acids, amino acids, and parts of the nucleotides also feed into various points in the TCA cycle to completely oxidize the carbon backbones, or to provide intermediates for other synthetic pathways. This cycle occurs when *oxygen is present* because all of the reducing power needs to go to the electron transport chain for ATP production, as well as the regeneration of the oxidized forms of the coenzymes (NAD^+ and FAD).

The TCA cycle has several important purposes (*why*). The TCA cycle is responsible for one of the main mechanisms of excreting carbons as CO_2, with two molecules of CO_2 formed per turn of the cycle. One turn of the TCA cycle also yields one GTP, which is equivalent in energy to an ATP, and four molecules of reducing power (three NADH and one $FADH_2$). Some of the TCA cycle intermediates can also be used for anabolic (synthetic) processes. Thus, the TCA cycle is considered an amphibolic pathway, meaning it serves both catabolic and anabolic functions. The TCA cycle takes place in the mitochondrial matrix (*where*), though the succinate dehydrogenase enzyme of the TCA cycle is actually part of Complex II of the electron transport chain in the inner mitochondrial membrane.

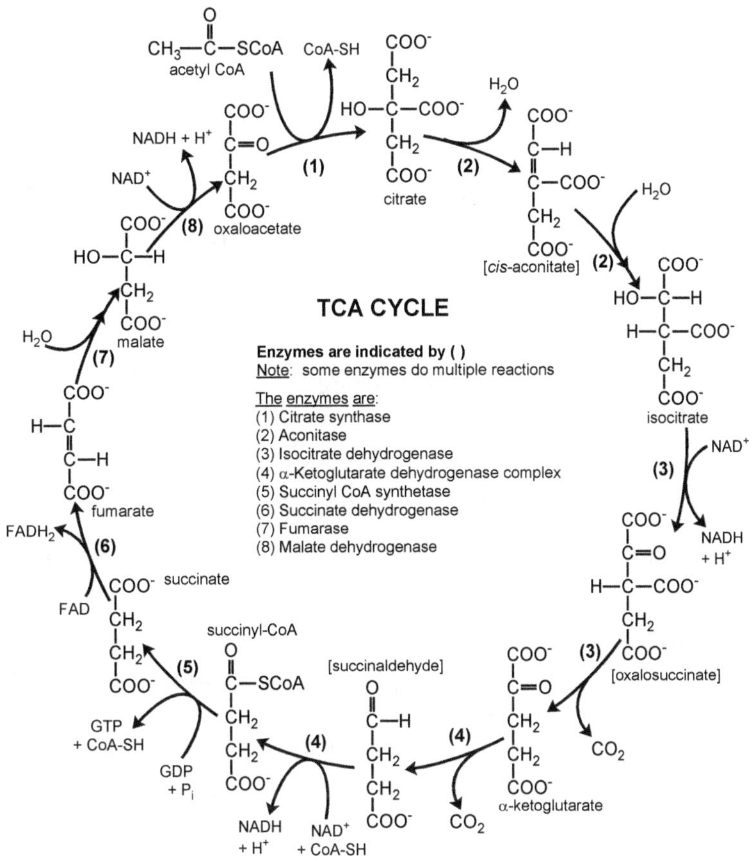

Figure 3.6: The TCA cycle.

The complete TCA cycle (*how*) is shown in **figure 3.6**, but each step will be covered individually. The cycle is divided into eight steps based on its eight enzymes. Some of the enzymes carry out more than one reaction, though. After the steps have been covered individually, one should come back to look at the complete cycle to walk through the entire cycle or draw it out separately to ensure understanding of the reaction sequence, enzyme names, and coenzymes needed. **Figure 3.6** does indicate both actual intermediates and some "fake" intermediates. All of the intermediates drawn are to emphasize how the pathway follows the oxidation states flowchart (**figure 1.7**). Always know starting and ending products, which for this pathway are acetyl CoA, oxaloacetate, and citrate. The rationales presented for these steps are *not* explaining the actual enzyme mechanisms, just emphasizing the pattern of the series of reactions.

In the TCA cycle several of the enzymes carry out multiple reactions. Key intermediates in these enzymatic reactions are indicated by [ ]. These intermediates are specifically drawn, so correlations to the oxidation states flowchart (**figure 1.7**) and the stability of carboxylic acids (**figure 1.14**) can be more readily seen.

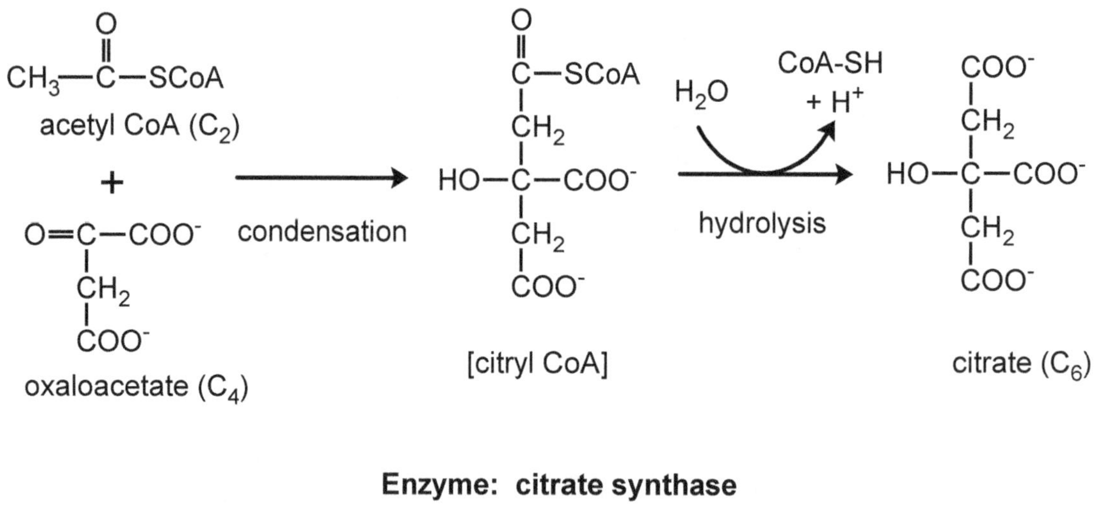

Enzyme: citrate synthase

Figure 3.7: Step 1: Entry of acetyl CoA into the TCA cycle by condensation with oxaloacetate.

Figure 3.7 shows the condensation reaction between the four-carbon oxaloacetate and the two-carbon unit of acetyl CoA to make the six-carbon citrate. Citrate is also known as tricarboxylic acid, due to the three carboxylic acid groups on it, giving the pathway its name. The molecule of oxaloacetate is drawn "bent" in **figure 3.7** to emphasize which carbons become attached between the two molecules. The carbonyl carbon of oxaloacetate forms a covalent bond to the methyl carbon of acetyl CoA, such that the entire acetyl CoA molecule is attached to oxaloacetate. This forms the intermediate citryl CoA. Water is used to cleave off coenzyme A, which now has a free sulfhydryl group or thiol (–SH) group. This hydrolysis of citryl CoA pulls the overall reaction far in the direction of forming citrate. Note that citrate can also be used as the transporter of the acetyl unit to the cytosol for fatty acid synthesis, as will be discussed when covering that pathway in chapter 7.

This is a synthesis reaction in a catabolic pathway. Enzymes that synthesize molecules are either named as a *synthase* (no energy required) or *synthetase* (energy required). Then place the name of the *product* before "synthase" or "synthetase." Since no energy is required in this reaction, the enzyme is called *citrate synthase*.

In the five-carbon "straight chain" part of citrate, notice the middle carbon is attached to *three* carbons and a hydroxyl group. Thus, this part of the molecule is a tertiary alcohol. This tertiary alcohol *cannot* be further oxidized because one of the carbon-carbon bonds would have to be broken so that two bonds can be made to the oxygen. This cannot be done unless they have good leaving groups attached. While one of the groups attached to this center carbon is a carboxylic acid, there must be a keto group nearby to get it to leave, which is not there—yet.

Citrate does have three carboxylic acid groups on it, and one goal of the TCA cycle is to remove two of them. In biological molecules the removal of carboxylic acid groups as molecules of CO_2 requires a keto group alpha (α) or beta (β) to it. There is a hydroxy group on the molecule of citrate that could become

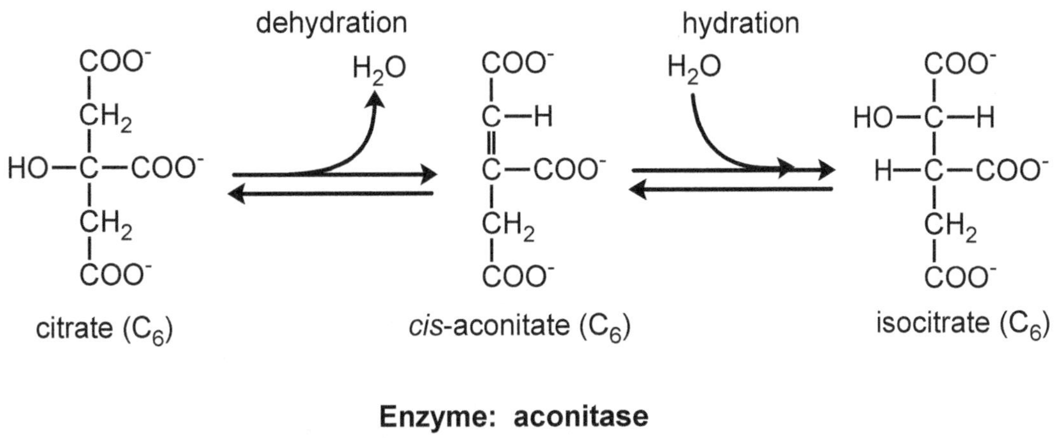

Figure 3.8: Step 2: Isomerization of citrate to isocitrate.

a keto group if it was a secondary alcohol, rather than the tertiary alcohol it is currently. Therefore, to be able to oxidize this hydroxy group to a keto group, it needs to be moved off of that center carbon of citrate.

Figure 3.8 shows the reaction sequence catalyzed by *aconitase* that "moves" the hydroxy group. Aconitase moves the hydroxy group first by a *dehydration* reaction to make the alkene intermediate *cis*-aconitate, followed by a *hydration* reaction where the hydroxy (–OH) group is now on one of the "original" CH_2 carbons forming the product isocitrate. Recall from the oxidation states flowchart (**figure 1.7**) that alkenes and alcohols are at the same oxidation state and can be interchanged by the addition or removal of an entire water molecule. This enzyme does that exact interchange, only now between a tertiary alcohol, an alkene, and a secondary alcohol. The hydroxy group is now a secondary alcohol on isocitrate and can be subsequently oxidized to a ketone.

As hydration and dehydration reactions are often reversible, the names of enzymes that carry out these reactions often seem to be varied (recall enolase from glycolysis) rather than following specific rules. Enzymes that specifically carry out hydration reactions are typically called *hydratases* named for the molecule accepting the water molecule. Enzymes that catalyze dehydration reactions are typically referred to as *dehydratases* and are named for the molecule losing the water molecule. However, this enzyme carries out both a dehydration and hydration reaction. Thus, the enzyme *aconitase* is simply named after the intermediate in this reaction, *cis*-aconitate.

Figure 3.9 shows the two reactions carried out by the enzyme *isocitrate dehydrogenase*, which catalyzes the first of four redox reactions of the TCA cycle. Isocitrate is first oxidized and then decarboxylated to form α-ketoglutarate. The first reaction is *oxidation* of the secondary alcohol group on isocitrate to a keto group to form the intermediate oxalosuccinate. If oxidation occurs, a molecule must be reduced. Since this oxidation is from an alcohol to a ketone, NAD^+ is reduced to $NADH + H^+$.

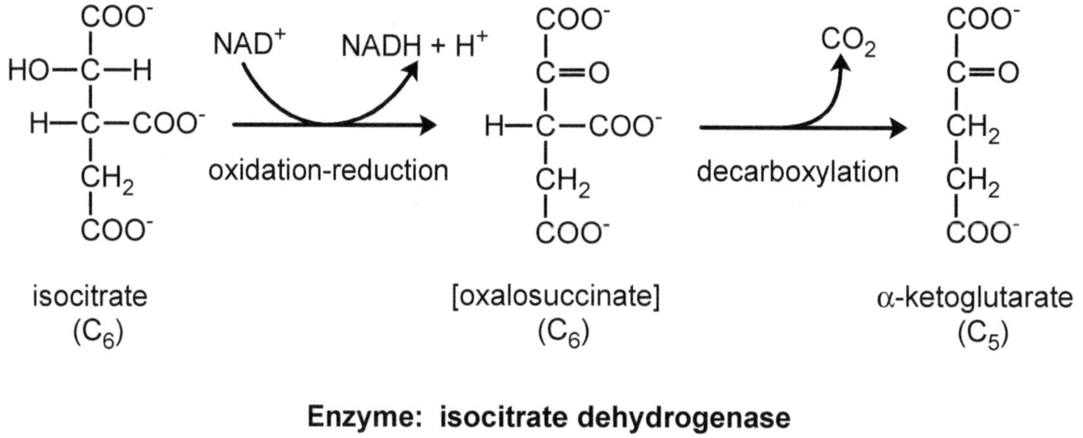

Enzyme: isocitrate dehydrogenase

Figure 3.9: Step 3: The first oxidation-reduction reaction of the cycle.

Notice the position of this keto group on oxalosuccinate relative to two out of the three carboxylic acid groups. The newly formed keto group is alpha (α) to one of the carboxylic acid groups (the top one in **figure 3.9**) and beta (β) to a second carboxylic acid group (the one on the center carbon of the molecule). A carboxylic acid that is located beta to a keto group is more unstable. Thus, the carboxylic acid on the center carbon of oxalosuccinate will leave first.

The second reaction carried out by isocitrate dehydrogenase is *decarboxylation* (loss of CO_2) of that β-keto carboxylic acid group to produce α-ketoglutarate, which is five carbons. While β-keto acids are unstable and spontaneously decarboxylate, "spontaneous" does not mean "instantaneous." Spontaneous does not indicate how long it will take for the carboxylic acid group to leave. Therefore, enzymes still typically carry out the decarboxylation of β-keto acids. Anytime an enzyme carries out more than one reaction, if one of the reactions is an oxidation-reduction reaction and uses NAD^+ or FAD, the enzyme is called *dehydrogenase*. Dehydrogenases are named for the *more reduced* molecule, so this enzyme is called *isocitrate dehydrogenase*.

The rate of α-ketoglutarate formation is important in determining the overall rate of the TCA cycle. The product, α-ketoglutarate, is also an amino acid precursor (or breakdown product) and can be converted to glutamate by a transamination reaction (**figure 1.5**).

The α-ketoglutarate dehydrogenase complex reaction series shown in **figure 3.10** is the simplified two-step reaction sequence that is exactly analogous to the PDH complex two-step mechanism shown in **figure 3.4**. This is an enzyme complex consisting of three catalytic enzymes. The complex carries out multiple reactions because the removal of an α-keto acid is difficult, as seen for the decarboxylation of pyruvate. This complex is a regulated step of the TCA cycle. However, it does not have the kinase and phosphatase regulatory subunits like the PDH complex.

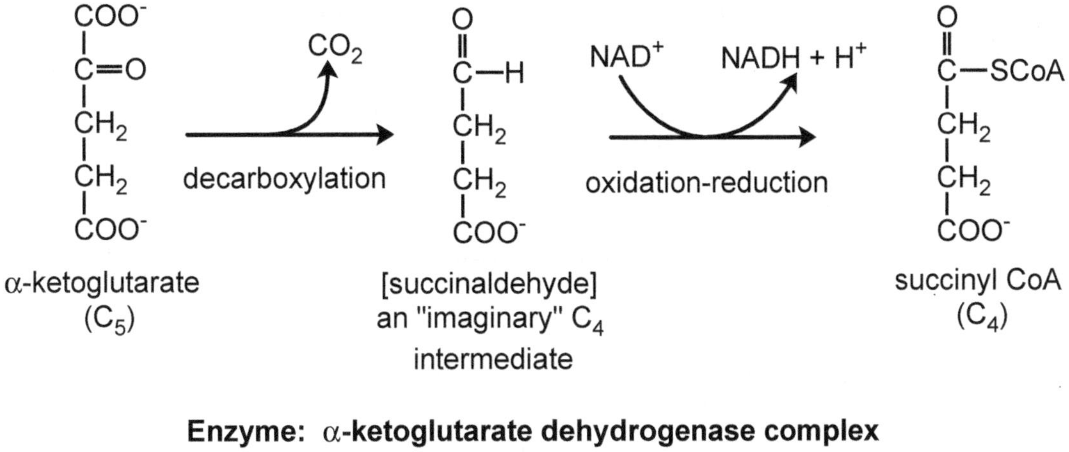

$$\underset{\substack{\alpha\text{-ketoglutarate} \\ (C_5)}}{\begin{array}{c} COO^- \\ | \\ C=O \\ | \\ CH_2 \\ | \\ CH_2 \\ | \\ COO^- \end{array}} \xrightarrow[\text{decarboxylation}]{CO_2 \uparrow} \underset{\substack{\text{[succinaldehyde]} \\ \text{an "imaginary" } C_4 \\ \text{intermediate}}}{\begin{array}{c} O \\ \| \\ C-H \\ | \\ CH_2 \\ | \\ CH_2 \\ | \\ COO^- \end{array}} \xrightarrow[\text{oxidation-reduction}]{NAD^+ \quad NADH + H^+} \underset{\substack{\text{succinyl CoA} \\ (C_4)}}{\begin{array}{c} O \\ \| \\ C-SCoA \\ | \\ CH_2 \\ | \\ CH_2 \\ | \\ COO^- \end{array}}$$

Enzyme: α-ketoglutarate dehydrogenase complex

Figure 3.10: Step 4: The second redox reaction of the cycle. Shown as a simplified two-step reaction sequence of the α-ketoglutarate dehydrogenase complex (as applies to the oxidation states flowchart).

The complex also does a decarboxylation reaction and an oxidation-reduction reaction, like isocitrate dehydrogenase, only in the reverse order. This time *decarboxylation* happens first to yield the intermediate succinaldehyde (a fake intermediate), which is four carbons. Then *oxidation* takes place, ultimately by the addition of coenzyme A to form a thioester group, which is equivalent to the carboxylic acid group oxidation state. So the molecule has been oxidized from a four-carbon aldehyde to the four-carbon succinyl CoA. As oxidation must be coupled with reduction, NAD$^+$ is reduced to NADH + H$^+$ when oxidizing an aldehyde to a carboxylic acid.

For the naming of this enzyme complex, multiple reactions are being carried out. One of the reactions is an oxidation-reduction reaction using NAD$^+$ or FAD, so the enzyme complex is called a *dehydrogenase*. Dehydrogenases are named for the *more reduced* molecule. In this case the more reduced molecule is an intermediate that cannot be isolated, so the enzyme is named the α-ketoglutarate dehydrogenase complex for the initial substrate: α-ketoglutarate. (Recall that the PDH complex is named in a similar manner.)

The actual reaction mechanism for removing an α-keto carboxylic acid group is very similar to the PDH complex, complete with final addition of CoA to the molecule (just like forming acetyl CoA). It requires three catalytic enzymes, which have the same prosthetic groups: thiamine pyrophosphate, lipoamide, and FAD. This reaction is very favorable and irreversible ($\Delta G° = -8.0$ kcal/mol). The products of the complex are CO_2, NADH, and succinyl CoA. Carbon dioxide is excreted, NADH is a substrate for the electron transport chain, and succinyl CoA can be used for heme biosynthesis.

At this point in the TCA cycle, succinyl CoA is a four-carbon molecule. Of the original three carboxylic acid groups that were on the six-carbon citrate, two have been removed as CO_2. Since the TCA cycle ends with the formation of the four-carbon oxaloacetate, no more carbons will be lost in the remaining reactions of the cycle.

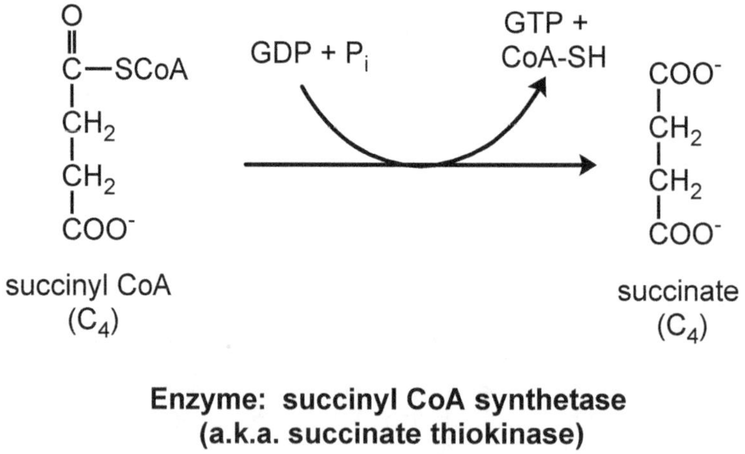

**Enzyme: succinyl CoA synthetase
(a.k.a. succinate thiokinase)**

Figure 3.11: Step 5: A substrate-level phosphorylation, which produces GTP.

Figure 3.11 shows the reaction carried out by *succinyl CoA synthetase*. This reaction is a *substrate-level phosphorylation*, where the phosphorylation of GDP to make GTP occurs at the expense of a high-energy organic substrate, succinyl CoA. The thioester bond of succinyl CoA is a high-energy bond ($\Delta G°'$ for the thioester hydrolysis is approximately -8.6 kcal/mol). Cleavage of this bond provides enough energy to drive the phosphorylation of GDP + P_i to make GTP (the phosphoanhydride bond created has approximately -7.3 kcal/mol). Note that succinyl CoA has *no* phosphate groups on it. Inorganic phosphate (P_i) is the donor of the phosphate. The product of this reaction is the four-carbon molecule succinate, and this reaction is readily reversible ($\Delta G°' = -0.7$ kcal/mol).

This enzyme is known as *succinyl CoA synthetase* or as *succinate thiokinase*. Both names of this enzyme are based on the *reverse* reaction, as if the energy from GTP is used to add coenzyme A to succinate to form succinyl CoA. For the name *succinyl CoA synthetase*, this enzyme is named as if succinyl CoA is being synthesized and the enzyme naming rules are more obviously applied. If you consider the reverse reaction, it would be a synthesis reaction that requires the use of energy. GTP is an energy equivalent to ATP. Thus, name the enzyme a *synthetase* and use the name of the product. Considering the reverse reaction, the product would be succinyl CoA. Therefore, the enzyme is named *succinyl CoA synthetase*.

In naming the enzyme *succinate thiokinase*, the rules for naming kinases still apply. This enzyme is being named for the *reverse* reaction. In this case GTP is "donating" the *energy* this time (*not* the actual phosphate group) from cleaving the high-energy phosphate bond to creating the thioester bond of succinyl CoA. Or more simply, a *thiokinase* is donating a thiol group to the acceptor molecule rather than the usual phosphate group of a typical kinase. Thus, succinate is still the acceptor of both the thiol group from coenzyme CoA and the energy from cleaving the phosphate bond to make it. Hence, the enzyme for this reaction is also called *succinate thiokinase*.

This is the only step of the TCA cycle that directly produces a molecule of chemical energy, in the form of GTP. The name succinyl CoA synthetase, in the author's opinion, is a better name to help one remember that energy is produced in this step. GTP is, like ATP, a form of chemical energy used by cells. GTP is often used for protein synthesis, signal transduction processes (i.e., G protein–coupled receptors), and the γ-phosphate group can be transferred to ADP to form ATP by a nucleoside diphosphate kinase.

*The last three reactions regenerate oxaloacetate (C_4) so that the cycle can continue. At this point, look at the differences in structure between the four-carbon succinate molecule (**figure 3.11**) and the four-carbon oxaloacetate molecule. Notice that both have carboxylic acid groups at each end of the molecules. However, the two middle carbons of succinate are at the level of an alkane, while one of the middle carbons of oxaloacetate has a keto group. To put that keto group on the molecule, the same sequence will be used as outlined in the oxidation states flowchart (**figure 1.7**): (1) oxidize the alkane to an alkene, (2) hydrate to form a secondary alcohol, and (3) oxidize to a ketone.*

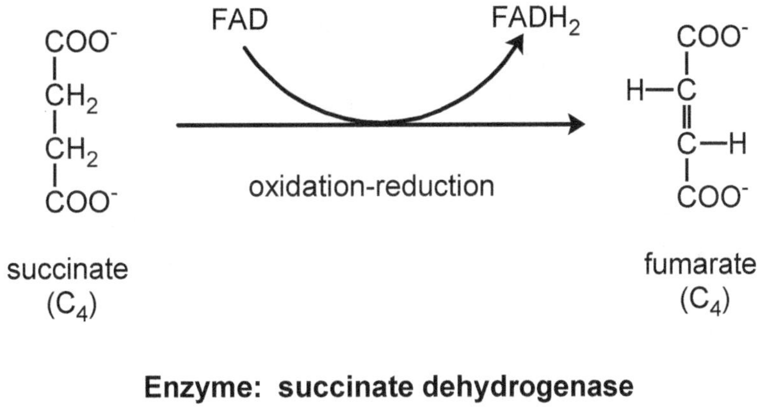

Enzyme: succinate dehydrogenase

Figure 3.12: Step 6: The third redox reaction of the cycle.

Succinate dehydrogenase catalyzes the oxidation-reduction reaction shown in **figure 3.12**. The two middle carbons of succinate are at the level of an alkane and are oxidized to the alkene of fumarate. Since the oxidation is occurring from the level of an alkane to an alkene, FAD is the acceptor of the hydrogens and is reduced to $FADH_2$. FAD is used here because the free energy change (ΔG) is not sufficient to reduce NAD^+. Enzymes that carry out oxidation-reduction reactions (and use NAD^+ or FAD) are called *dehydrogenases*. Dehydrogenases are named for the *more reduced* molecule, which for this reaction is succinate. Thus, the enzyme is called *succinate dehydrogenase*.

The FAD used in this reaction is covalently attached to a histidine ring of the enzyme, meaning that it cannot dissociate from the enzyme. Succinate dehydrogenase, though, is actually part of Complex II of

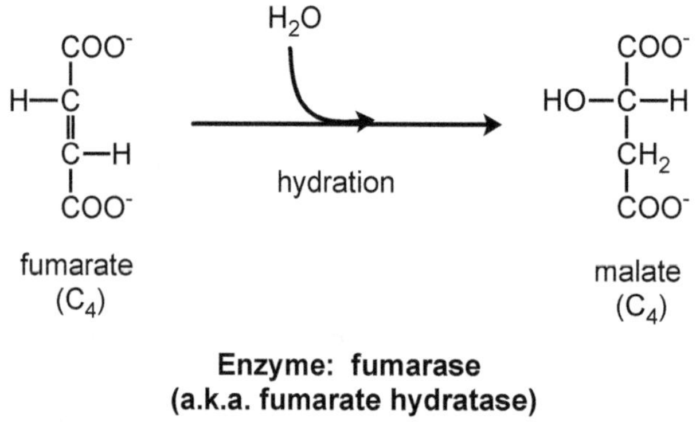

```
         COO⁻                H₂O                  COO⁻
          |                                        |
        H—C                                     HO—C—H
          ‖           ──────────►                  |
        C—H                                       CH₂
          |              hydration                 |
         COO⁻                                     COO⁻

        fumarate                                  malate
          (C₄)                                     (C₄)
```

Enzyme: fumarase
(a.k.a. fumarate hydratase)

Figure 3.13: Step 7: A hydration step to go from an alkene to a secondary alcohol.

the electron transport chain. Therefore, the electrons of the $FADH_2$ formed in this reaction are transferred immediately to the electron transport chain.

Figure 3.13 shows the hydration reaction carried out by *fumarate hydratase*, which is more commonly known as *fumarase*. Fumarate is hydrated, the addition of an entire water molecule across the double bond, to form the secondary alcohol group on malate. Recall that alkenes and alcohols are at the same oxidation level, so this is not a redox reaction. If using the enzyme naming rules, *hydratases* are enzymes that add water to a molecule. Hydratases are typically named for the molecule to which the water molecule is added. In this case, *fumarate hydratase* is the more accurate enzyme name based on naming rules.

The last step of the cycle is the oxidation of malate (secondary alcohol group) to oxaloacetate (keto group), shown in **figure 3.14**. Since the oxidation is occurring from a secondary alcohol to a ketone, NAD^+ is reduced to $NADH + H^+$. Again, enzymes that carry out redox reactions and use NAD^+ or FAD are called *dehydrogenases*. Dehydrogenases are named for the *more reduced* molecule, which is malate for this reaction. Thus, the enzyme is called *malate dehydrogenase*.

The reaction for the formation of oxaloacetate is very endergonic and unfavorable. The $\Delta G°'$ for this reaction, as written for the formation of oxaloacetate, is +7 kcal/mol. Thus, the reverse reaction, the formation

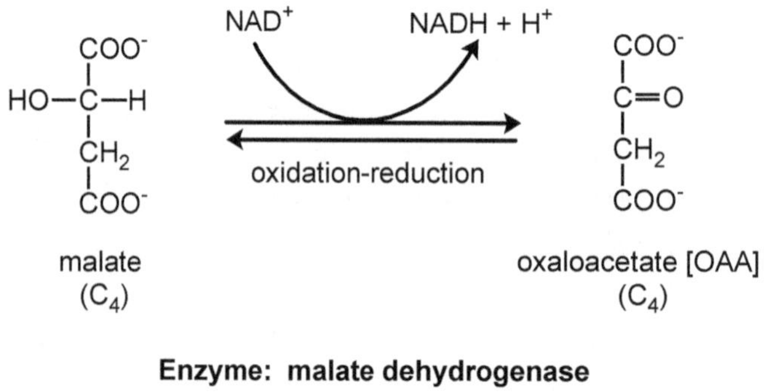

```
                         NAD⁺        NADH + H⁺
         COO⁻                                        COO⁻
          |                ↘         ↗                |
        HO—C—H                                       C=O
          |             ──────────►                   |
         CH₂            ◄──────────                  CH₂
          |             oxidation-reduction           |
         COO⁻                                        COO⁻

        malate                                oxaloacetate [OAA]
         (C₄)                                       (C₄)
```

Enzyme: malate dehydrogenase

Figure 3.14: Step 8: The fourth redox reaction completes the cycle.

of malate, has a ΔG°' of –7 kcal/mol. The formation of malate by malate dehydrogenase should then be essentially an irreversible reaction, but it is not. Cells bypass irreversible reactions (if necessary) typically in one of two ways. The reverse reaction is a different reaction and uses different enzymes, as seen in gluconeogenesis for bypassing the three irreversible reactions of glycolysis. Or a reaction is coupled to a favorable reaction to make a reaction go in an unfavored direction, as free energies (ΔG) of reactions are additive. In this case oxaloacetate is the product of one reaction and the substrate of the subsequent reaction. The key is to keep the malate concentration high and the oxaloacetate concentration in the mitochondria very low (i.e., as soon as it is formed, it is used) by its immediate use in the citrate synthase reaction, which is very favorable (ΔG°' = –7.7 kcal/mol). Thus, malate dehydrogenase continues to form oxaloacetate under these conditions.

Another key determinant in regulating which direction the malate dehydrogenase reaction goes is the ratio of NADH/NAD$^+$ in the mitochondria. When NADH is high in the mitochondria, typically the electron transport chain is slowing down or stopping because plenty of ATP has been made. This results in the buildup of NADH and FADH$_2$, which inhibits the TCA cycle. These conditions would allow for activation of gluconeogenesis, a synthesis pathway, resulting in the reduction of oxaloacetate to malate by malate dehydrogenase for transport to the cytosol.

Regulation of the TCA cycle and biosynthetic uses of its intermediates

Citrate synthase, isocitrate dehydrogenase, and the α-ketoglutarate dehydrogenase complex are the three TCA cycle enzymes that are regulated. While each of these enzymes has several inhibitors and activators, the focus here is on ATP availability as a basic concept of regulation of any metabolic pathway. One should consider if high concentrations of ATP (i.e., a high ATP/ADP ratio) would activate or inhibit the TCA cycle. If the cell has plenty of ATP, the electron transport chain would be inhibited. High levels of ATP and NADH would result in the mitochondria, which would inhibit the TCA cycle. NADH inhibits all three of these regulated enzymes, while ATP is an inhibitor of both citrate synthase and isocitrate dehydrogenase. The energy state of the cell also regulates the α-ketoglutarate dehydrogenase complex, since AMP activates the complex.

As mentioned earlier, the TCA cycle is a provider of intermediates for other biosynthetic pathways. Succinyl CoA provides carbon atoms for porphyrin (heme) synthesis, and α-ketoglutarate and oxaloacetate are intermediates for several amino acid syntheses. The problem is that TCA cycle intermediates must be replenished if any are drawn off for biosynthetic reactions and the cycle needs to continue. The solution is that other reactions, termed anaplerotic reactions (meaning to "fill up"), replenish intermediates. The pyruvate carboxylase reaction can replenish oxaloacetate, and the catabolism of several amino acids form intermediates such as acetyl CoA, fumarate, α-ketoglutarate, and oxaloacetate.

NET ATP YIELD FROM COMPLETE CATABOLISM OF GLUCOSE

Table 3.2 shows the calculation of net ATP yield from the complete catabolism of a glucose molecule to six molecules of CO$_2$ (and water). The catabolism of glucose includes both molecules of pyruvate from glycolysis yielding two molecules of acetyl CoA, via the PDH complex, to turn the TCA cycle two times. To calculate

Table 3.2: ATP yield from the complete catabolism of a molecule of glucose

LOCATION	PATHWAY	YIELD/GLUCOSE	ATP
Cytoplasm	Glycolysis	–2 ATP 4 ATP 2 NADH	–2 +4 +5 (or +3)*
Mitochondria	PDH complex	2 NADH	+5
Mitochondria	TCA cycle	2 GTP 6 NADH 2 $FADH_2$	+2 +15 +3
Total			32 (or 30)* ATP

*Numbers in brackets indicate lower ATP yield for the glycerol phosphate shuttle, versus the higher number if the malate-aspartate shuttle is used.

the ATP yield, one needs to consider how the reducing power (NADH and $FADH_2$) produced from glycolysis, the PDH complex, and the TCA cycle is used by the electron transport chain to make ATP. While the electron transport chain will be discussed in chapter 4, the general outcome of the electron transport chain is mentioned here. NADH donates two electrons to the electron transport chain at Complex I, while $FADH_2$ donates its two electrons at Complex II. The electron movement from NADH through Complexes I, III, and IV to ultimately end up on oxygen, reducing it to water, produces enough electrical energy (proton gradient) to drive the synthesis of about 2.5 ATP (chemical energy). The amount of ATP produced from the reduction of oxygen is termed the P/O ratio. The electron movement from $FADH_2$ through Complexes II, III, and IV to ultimately end up reducing molecular oxygen to water produces enough electrical energy to drive the synthesis of about 1.5 ATP. So the P/O ratio for NADH is 2.5 ATP per oxygen consumed, and $FADH_2$ is 1.5 ATP per oxygen consumed.

In calculating the total yield of ATP from glucose, one must also consider which shuttle is used to transport the electrons from the NADH produced in the cytosol by glyceraldehyde 3-phosphate dehydrogenase in glycolysis. If the malate-aspartate shuttle is used, the electrons from the cytosol end up on a mitochondrial NADH molecule. If the glycerol phosphate shuttle is used, the cytosolic electrons end up on a molecule of $FADH_2$ in the mitochondria, which would yield less ATP. Two molecules of ATP are also used in the beginning of glycolysis, so these will be subtracted in calculating the net yield.

The complete catabolism of a glucose molecule will yield about 30 to 32 ATP for the cell. In a bomb calorimeter, the combustion of glucose to six CO_2 and six H_2O would yield approximately 2,870 kJ (686 kcal) of energy released as heat. The hydrolysis of one ATP (to ADP + P_i) yields –7.3 kcal/mol. The hydrolysis of 32 ATP yields –234 kcal/mol. The efficiency of the cell to harness the energy released from the oxidation of glucose is about 34 percent (–234 kcal/mol divided by –686 kcal/mol), which is actually reasonably efficient. To put this number in perspective, most automobile engines are not this efficient in harnessing the energy from the combustion of gasoline.

IMAGE CREDIT

- Fig. 3.1: Mariana Ruiz Villarreal, "Basic Structure of a Mitochondrion," https://commons.wikimedia.org/wiki/File:Animal_mitochondrion_diagram_en.svg. Copyright in the Public Domain.

CHAPTER 4

THE ELECTRON TRANSPORT CHAIN

OBJECTIVES

1. Define the process of the electron transport chain.
2. Explain the purpose of the electron transport chain, including the special function of it in brown adipose tissue.
3. Identify where the electron transport chain takes place in the cell.
4. Explain how the electron transport chain is carried out in the cell and in specialized tissues (i.e., brown adipose tissue).
   a. Identify the two types of prosthetic groups that carry both protons and electrons, and the three types of prosthetic groups that only carry electrons.
   b. Identify the specific electron donors (substrates) and final electron acceptor (product) of each of the complexes, I through IV.
   c. Identify which of the complexes (I through IV) is capable of pumping protons.
5. Explain when the electron transport chain occurs and describe the chemiosmotic mechanism of oxidative phosphorylation.
   a. Explain how the membrane potential is generated and used.
   b. Describe the mechanism of uncoupling the membrane potential from ATP generation and why this would be necessary.
   c. Define the term respiratory control.
6. Describe the special features of brown adipose tissue mitochondria as generators of heat.

THE ELECTRON TRANSPORT CHAIN

The electron transport chain is extremely complex. This chapter will present a simplified, or big picture, view of the electron transport chain. The focus will be placed on the electron donors and acceptors of each complex, and the coupling (or uncoupling) of the proton gradient with ATP synthase. A basic explanation will be provided regarding how the various prosthetic groups associated with Complexes I through IV move electrons, while pumping protons from the mitochondrial matrix to the intermembrane space.

The electron transport chain (*what*) is a series of oxidation-reduction reactions using NADH and $FADH_2$ as the initial electron donors, which is why they are termed "reducing power," and molecular oxygen as the terminal electron acceptor. The movement of electrons through Complexes I through IV will produce a proton gradient. The proton gradient will be coupled by Complex V (ATP synthase) to the phosphorylation of ADP to make ATP. The coupling of electron movement to ATP synthesis is called *oxidative phosphorylation*.

The purpose (*why*) of the electron transport chain is usually to produce ATP. Energy that is not captured in the synthesis of ATP is primarily lost as heat, which is used to maintain body temperature. Brown adipose tissue, also known as brown fat, is a specialized tissue that actually produces heat rather than ATP, as will be further discussed. The five complexes of the electron transport chain are located in the inner mitochondrial membrane (*where*).

Figure 4.1 shows the overall energy needed to synthesis three molecules of ATP (+22 kcal/mol). The oxidation of NADH yields −38 kcal/mol, and the reduction of one oxygen atom from molecular oxygen to water yields −14 kcal/mol. The coupling of these two reactions yields a total $\Delta G^{o\prime}$ of −52 kcal/mol. These two coupled reactions certainly provide enough energy to synthesis three molecules of ATP (i.e., more than the +22 kcal/mol needed). Coupling of these reactions only needs to be about 40 percent efficient (i.e., 22/52) as the remaining energy is lost as heat, which is used to maintain body temperature. The oxidation of $FADH_2$ coupled to the reduction of oxygen to water generates approximately −40 kcal/mol. An approximate 40 percent efficiency of these two reactions would provide enough energy for the synthesis of about two ATP.

The ratio of three ATP produced per oxygen consumed (the P/O ratio) from NADH and two ATP per oxygen consumed from $FADH_2$ are still common in biochemical literature, as the initial theories on how the electron transport chain worked indicated this ratio should always be an integer. However, the proposal

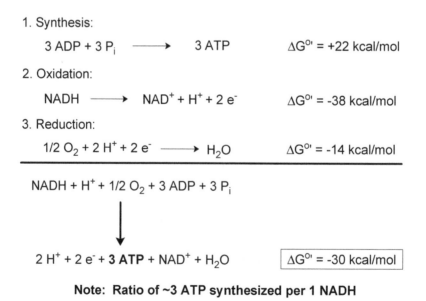

Figure 4.1: Overall process of the electron transport chain.

that a membrane potential was what coupled electron movement to ATP synthesis meant that the P/O ratio did not need to be an integer. Further research lead to consensus values for the number of protons pumped per electron pair from the oxidation of NADH (ten H$^+$) and succinate/FADH$_2$ (six H$^+$). The generally accepted number of protons needed to drive ATP synthesis is four, with one of these protons needed for the transport of ADP, P$_i$, and ATP across the inner membrane. For NADH, 10 H$^+$ divided by 4 H$^+$/ATP is 2.5 ATP per oxygen consumed. For succinate/FADH$_2$, 6 H$^+$ divided by 4 H$^+$/ATP is 1.5 ATP per oxygen consumed. Thus, the accepted values are *2.5 ATP per NADH and 1.5 ATP per FADH$_2$* can be synthesized.

Obviously, there is plenty of energy to get the job done, but how are these reactions coupled? Metabolic energy from oxidation of food materials (sugars, fats, and amino acids) is funneled into the formation of reducing equivalents: NADH and FADH$_2$. NADH and FADH$_2$ donate their electrons to Complexes I and II, respectively. These electrons move through the complexes, ultimately reducing molecular oxygen to water. As represented in **figure 4.2**, the movement of electrons (reaction 1) through the complexes provides energy to move protons (H$^+$) from the matrix to the intermembrane space forming a membrane potential (~200 mV), which is electrical energy. The membrane potential (*electrical energy*) is then used by Complex V (ATP synthase) in reaction 2 to drive ATP synthesis, which is the primary form of *chemical energy* used by cells. Peter Mitchell won the Nobel Prize in Chemistry in 1978 for his theory that the mitochondria establish and utilize a membrane potential to drive ATP synthesis, called the chemiosmotic principle.

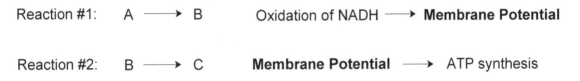

Figure 4.2: Overview of a coupled electron transport chain.

To further illustrate how the electron transport chain works, think of the analogy of pushing a large rock up a hill. The energy spent pushing this large rock up the hill is now stored as potential energy as the rock sits on the hill. If the rock was then pushed off the hill and allowed to roll back to the bottom, the energy would simply be lost as heat, and no useful work would be done. Now envision a rope attached to this large rock that has once again been pushed to the top of the hill. This rope is run through a pulley system and attached to a heavy crate loaded with supplies at the bottom of the hill. Now when the large rock is pushed off the top of the hill, it pulls the rope, which in turn lifts the load to the top of the hill. Now the potential energy (i.e., membrane potential) stored in the rock (i.e., the proton gradient) at the top of the hill has been converted to do useful work, lifting the crate (i.e., ATP synthesis).

Figure 4.3 shows a schematic representation of the electron transport chain. Complexes I through IV are a series of integral membrane proteins embedded in the inner membrane of the mitochondria. Complexes I through IV are responsible for the movement of the electrons and establishing the proton gradient. These four complexes have various prosthetic groups that accept and then pass on these electrons. Ultimately, the electrons end up on molecular oxygen, reducing it to two molecules of water. In the process of moving the

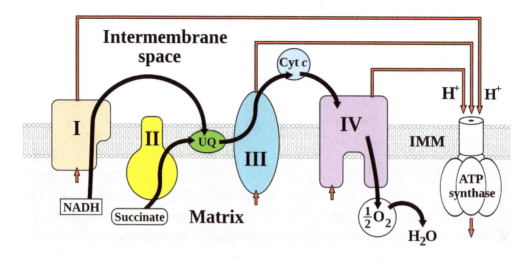

Figure 4.3: Schematic representation of the electron transport chain.

electrons, Complexes I, III, and IV are going pump protons (the red arrows in **figure 4.3**) from the matrix into the intermembrane space, generating the proton gradient.

Complex V is also an integral membrane protein. Complex V has a channel in the membrane portion that specifically allows protons to flow back from intermembrane space into the matrix. The release of the membrane potential by allowing the protons to flow back to the matrix is used by Complex V to synthesize ATP from ADP plus P_i.

How are the protons pumped across the inner mitochondrial membrane? Oxidation-reduction reactions, as discussed previously, move electrons. Recall the difference between a hydrogen atom (one proton + one electron), a hydride atom (one proton + two electrons), and a proton (one proton + no electrons). Up to this point, as humans are not made up of metal wires, the movement of electrons in redox reactions has occurred by the use of hydrogen atoms or hydride atoms. However, Complexes I through IV of the electron transport chain have several types of prosthetic groups that contain metal atoms, either iron or copper. Metals only want the electrons. If the metal atom of the prosthetic group only takes the electron from a hydrogen atom, what are you left with? Answer: a proton (H^+). As the electrons are passed from one prosthetic group to the next, they lose energy. The energy from the electron movement pumps the protons to the intermembrane space.

To summarize this big picture perspective, Complexes I through IV have various prosthetic groups. Some of the prosthetic groups accept full hydrogen atoms, while others have iron or copper atoms that only want the electrons. As the electron is removed from a hydrogen atom, the resulting proton is pumped to the intermembrane space (though some of the protons may be released back into the matrix). Overall, the net movement of protons is from the matrix to the intermembrane space as the electrons move through prosthetic groups that either want full hydrogen atoms or just electrons.

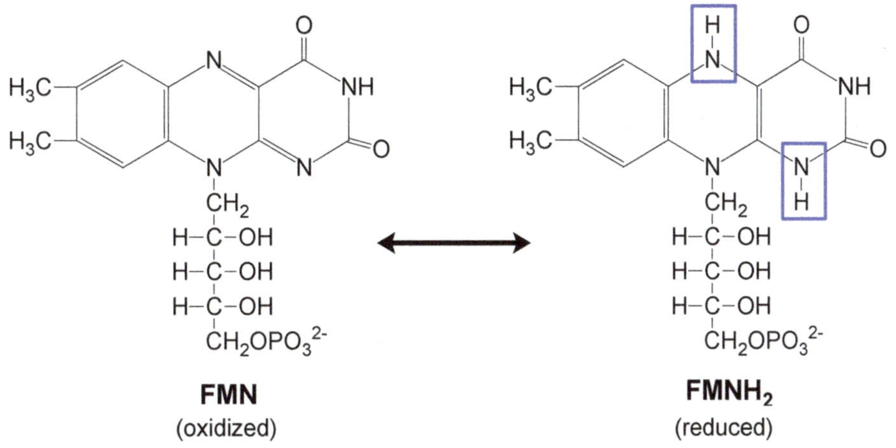

Figure 4.4: The structure of FMN (oxidized) and FMNH$_2$ (reduced).

Thus, there are two categories of prosthetic groups or coenzymes of the electron transport chain. The first category is prosthetic groups or coenzymes that accept full hydrogen atoms. Flavin adenine dinucleotide (FAD/FADH$_2$) has already been introduced. A related prosthetic group is flavin mononucleotide (FMN/FMNH$_2$). Both FAD and FMN are derived from the vitamin riboflavin (B$_2$) and have the same active site (hydrogen accepting sites). Both accept two full hydrogen atoms to form their reduced forms, FADH$_2$ and FMNH$_2$. The structure of FAD consists of FMN, shown in **figure 4.4**, with AMP attached at the phosphate group of FMN.

Coenzyme Q (CoQ) is another coenzyme of the electron transport chain that accepts two full hydrogen atoms. As shown in **figure 4.5**, CoQ has two keto groups in the oxidized form, which is called ubiquinone. The two keto groups are reduced to two hydroxy groups, forming ubiquinol, or CoQH$_2$. As shown in **figure 4.5**, CoQ has a very hydrophobic tail, consisting of ten isoprene units, which sequesters CoQ in the inner mitochondrial membrane. Rapid diffusion of CoQ through the inner membrane allows it to act as a carrier of electrons between Complexes I and III, as well as between Complexes II and III. CoQ is also involved in the electron transfer reactions of Complex III.

While FAD, FMN, and CoQ accept the whole hydrogen atom (proton + electron), there is a second category of prosthetic groups of the electron transport chain that only accept electrons. These are the heme groups (porphyrin rings plus iron), the iron-sulfur centers, and the copper groups. The heme groups are part of cytochromes, in which the iron atom in the middle of the heme is protected from oxygen. The electron transport chain contains three types of cytochromes, termed cytochromes *a*, *b*, and *c*. They are differentiated into these categories based on their light-absorption spectra. The heme cofactors of type *a* and *b* cytochromes tightly interact but are not covalently bound to the protein. Cytochromes of the *c* type do have the heme covalently bound to the protein. Complex III contains cytochromes *b* and c_1, while Complex IV contains cytochromes *a* and a_3. Cytochrome *c*, in contrast, is a soluble protein that remains associated with the outer surface of the inner membrane through electrostatic interactions. Note that **figure 4.3** does not accurately show the association of cytochrome *c* with the membrane. Cytochrome *c* serves as the carrier of electrons between Complexes III and IV.

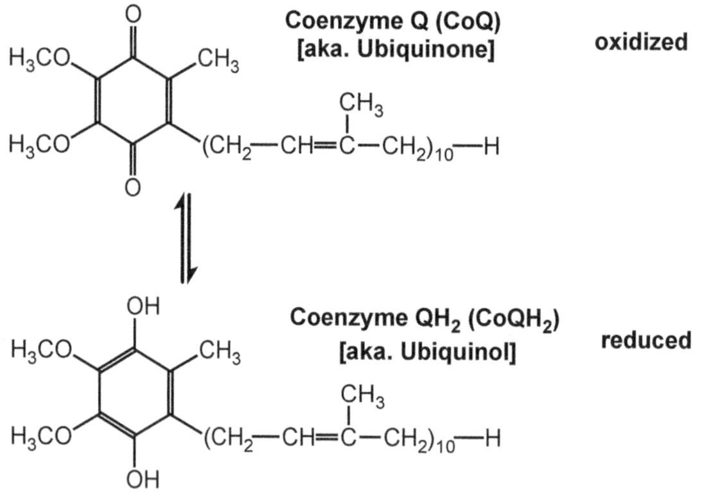

Figure 4.5: The oxidized and reduced structures of CoQ.

The iron atom in the center of the heme groups associated with any of the cytochromes accepts electrons only and alternates between the oxidized ferric state (Fe^{3+}) and the reduced ferrous state (Fe^{2+}). However, the standard reduction potential (E_o) of the iron atom of each cytochrome is different because it depends on how the heme interacts with its protein side chains. As oxidation-reduction reactions are always coupled, the electrons flow from the more negative E_o of the pair to the more positive E_o of the pair.

There are also various types of iron-sulfur centers located in Complexes I, II, and III. One example of an iron-sulfur center is a single iron atom coordinated to the sulfurs of four cysteine residues. Other iron-sulfur centers include multiple iron atoms coordinated to inorganic sulfur atoms and/or sulfurs from cysteine residues. The iron atoms of these iron-sulfur centers also alternate between the oxidized ferric state (Fe^{3+}) and the reduced ferrous state (Fe^{2+}) and only accept electrons.

Complex IV contains two copper atoms. The first, designated Cu_A, is complexed with the sulfhydryl groups of two cysteine residues. The second one is designated Cu_B. These copper ions, as metals, also only accept electrons.

The net movement of protons from the matrix to the intermembrane space is accomplished by passing the electrons through various prosthetic groups in the complexes that either accept whole hydrogen atoms (proton + electron) or only need the electrons and thus release a proton. The loss of energy as the electrons move through the complexes provides the necessary energy to pump protons from the matrix to the intermembrane space.

Each of the five complexes of the electron transport chain (shown in **figure 4.3**) will be reviewed briefly. For Complexes I through IV, three key aspects will be emphasized: the electron donor of the complex, the electron acceptor from the complex, and whether the complex is capable of pumping protons from the matrix to the intermembrane space. Some of the other prosthetic groups of the complexes will be noted, but additional details regarding the flow of electrons through the complexes will not be discussed. While the complexes are often simply referred to as Complex I, Complex II, and so on, they do have other names that will be noted. Certain names for these complexes actually indicate the electron donor and acceptor of the complex.

Complex I is known as *NADH-CoQ reductase*, *NADH dehydrogenase*, or *NADH-ubiquinone oxidoreductase*. The "reductase" or "oxidoreductase" names indicate the donor of electrons to the complex (*NADH*) and the acceptor of electrons from the complex (*ubiquinone, CoQ*). NADH donates two electrons to the complex as it is oxidized back to NAD^+. The electrons are then passed through various other prosthetic groups, including FMN and several iron-sulfur centers. CoQ is the final acceptor of the two electrons from Complex I, which reduces ubiquinone (CoQ) to ubiquinol ($CoQH_2$). The net result is that *protons are pumped* from the matrix to the intermembrane space as the electrons move through Complex I.

Complex II also has a couple of different names. Complex II is often simply referred to as *succinate dehydrogenase*, the same enzyme encountered in the TCA cycle. Recall that succinate dehydrogenase oxidized succinate to fumarate, while reducing FAD to $FADH_2$. As previously mentioned in the coverage of this reaction, succinate dehydrogenase is a component of Complex II of the electron transport chain. Other names for Complex II are *succinate-CoQ reductase* or *succinate-ubiquinone oxidoreductase*. These two names are more complete descriptions of the complex as they indicate the electron donor and acceptor of Complex II. As succinate dehydrogenase is part of this complex, succinate is often named as the donor of the electrons to the complex. However, it is important to know that Complex II is the point of entry of electrons from $FADH_2$. The FAD prosthetic group of succinate dehydrogenase does not leave the enzyme active site. Therefore, as soon as $FADH_2$ is formed as succinate is oxidized to fumarate, the $FADH_2$ can donate its electrons directly to Complex II and be re-oxidized back to FAD. Thus, one may consider succinate or $FADH_2$ as the electron donors to Complex II. The electrons move through several iron-sulfur centers before ending up on the final electron acceptor. The final electron acceptor of Complex II is another molecule of CoQ (ubiquinone), which is reduced to ubiquinol ($CoQH_2$). Both Complex I and Complex II give their two electrons to molecules of CoQ, which will then carry these electrons to Complex III. The movement of electrons through Complex II does not provide enough energy to pump protons across the membrane. Therefore, Complex II *does not pump protons*.

Complex III has many names, which include *cytochrome* c *reductase*, *ubiquinol-cytochrome* c *reductase*, *ubiquinone-cytochrome* c *oxidoreductase* or *cytochrome* bc_1 *complex*. The reduced form of CoQ ($CoQH_2$ or ubiquinol) carries the electrons from either Complex I or Complex II and donates two electrons to Complex III. As the electrons flow through this complex, they pass through cytochromes b and c_1, as well as through an iron-sulfur center. The final electron acceptor from Complex III is *oxidized* cytochrome *c*, which is a *one* electron carrier as it goes from its oxidized ferric state (Fe^{3+}) to its reduced ferrous state (Fe^{2+}), forming *reduced* cytochrome *c*. The mechanism of electron transfer in Complex III is complicated and involves what is termed the *Q cycle*. The complex mechanism is due to switching from the two-electron donor (ubiquinol) to a one-electron final acceptor, cytochrome *c*. Though the mechanism is intricate, the net result is that the two electrons donated by a molecule of ubiquinol ($CoQH_2$) end up reducing two molecules of cytochrome *c*. As the electrons move through the Complex III, there is enough energy generated to pump protons from the matrix to the intermembrane space. Thus, Complex III *does pump protons*.

Complex IV is called *cytochrome c oxidase*. Any time molecular oxygen is used in an oxidation-reduction reaction, the enzyme is called an *oxidase* (rather than a dehydrogenase or reductase). The complex is called *cytochrome c oxidase* because *reduced* cytochrome *c* is donating an electron to Complex IV and re-forms oxidized cytochrome *c* in the process. Molecular oxygen is the terminal electron acceptor of Complex IV and of the entire electron transport chain. As the electrons move through Complex IV, they also pass through cytochromes *a* and a_3, as well as two copper ions (Cu_A and Cu_B) described previously. The mechanism of electron movement through Complex IV is also complicated. Four electrons from four molecules of reduced cytochrome *c* are donated to the complex resulting in the reduction of *both* oxygen atoms of molecular oxygen to two water molecules. Partial reduction of oxygen would produce reactive oxygen species, which are detrimental to the cell. Thus, Complex IV does not release partial products. Only when both oxygen atoms are fully reduced are both water molecules released from the complex. Complex IV also *pumps protons* from the matrix to intermembrane space.

Molecular oxygen is the ideal terminal electron acceptor of the electron transport chain. It has a high affinity for electrons providing a large thermodynamic driving force. In other words, the coupled reactions of the oxidation of NADH and the reduction of oxygen to water is highly exergonic, as shown in **figure 4.1**. Molecular oxygen also reacts slowly, unless activated by a catalyst.

At this point, the movement of electrons through Complexes I through IV is done. By the end of the reactions of Complex IV, molecular oxygen is consumed by accepting four electrons until it is fully reduced to two molecules of water. The movement of electrons through Complexes I through IV has resulted in a net movement of protons (H^+) to the intermembrane space to create the necessary membrane potential that will be used by Complex V.

Complex V, the final complex of the electron transport chain (shown in **figure 4.3**), synthesizes ATP. Hence, the name of Complex V is *ATP synthase*. Note the name is ATP *synthase*, not synthetase. This complex is producing ATP, not using it. Thus, it is a synthase, rather than a synthetase. Complex V is also called the F_0F_1-*ATPase* because it has two distinct domains, F_0 and F_1. The F_0 domain contains a hydrophobic segment that allows it to reside in the inner membrane. This hydrophobic segment has a membrane pore or channel that allows protons to flow through it from the intermembrane space back into the matrix. The F_1 domain has numerous subunits, including three alpha (α) subunits, three beta (β) subunits, and a gamma (γ), delta (δ), and epsilon (ε) subunit. The three β subunits are the catalytic subunits, with each β subunit able to catalyze the reaction of ADP + P_i to yield ATP.

The net reaction of Complex V couples the membrane potential (~200 mV) plus ADP and P_i to yield ATP. How is this proton membrane potential used to drive ATP synthesis? Paul D. Boyer and John E. Walker both won the 1997 Nobel Prize in Chemistry for their work in this area. The mechanism is more complex, but in general terms, they determined that three to four protons flow through the channel (F_0) portion of Complex V, which induces a conformational change in F_1. They determined that the β subunits can catalyze the reaction of ADP + P_i to produce ATP without the proton gradient. However, the synthesized ATP remains in this catalytic subunit until the protons flow through to allow their release. ADP and P_i then cycle into the open slots, where they bind tightly to exclude water from the active site, allowing the phosphoanhydride bond to form easily. For the continued synthesis of ATP, the enzyme must cycle between a conformation that binds ATP tightly and a conformation that releases the ATP. The proton gradient allows for the cycling between these conformations.

In summary, the movement of electrons from NADH and succinate/FADH$_2$ to oxygen pumps protons into the intermembrane space to produce a proton gradient (i.e., membrane potential). By specifically controlling how these protons flow back to the matrix, via Complex V, the cell is able to convert an electrical energy form (the membrane potential) to a chemical energy form (ATP).

Regulation of the electron transport chain

As illustrated in other pathways, irreversible reactions are typically points of regulation for a pathway. The regulation of the electron transport chain occurs at Complex IV. The movement of the electrons through this complex to reduce oxygen to water is *irreversible*. This complex is controlled by the amount of reduced cytochrome *c*, the donor of the electrons to the complex. The reactions from NADH to the formation of reduced cytochrome *c* are readily reversible. Thus, reduced cytochrome *c* is in equilibrium with the earlier intermediates of the electron transport chain. The concentration of reduced cytochrome *c* depends on the ratios of NADH/NAD$^+$ and ATP/ADP + P$_i$, essentially the energy state of the cell. If there is plenty of ATP in the cell, Complex V will stop. The proton gradient will build up, and oxygen consumption by Complex IV will stop. Reduced cytochrome *c* will build up, causing the previous reactions to run in reverse to shift the equilibrium back toward the formation of NADH and FADH$_2$. The buildup of NADH and FADH$_2$ results in further inhibition of reactions in the TCA cycle and the PDH complex.

For a person whose is active, the hydrolysis of ATP is high. In an active person, the reduced cytochrome *c* concentrations (in appropriate cells) would increase, resulting in increased electron movement, oxygen consumption, and ATP synthesis. In an inactive person (i.e., at rest), ATP hydrolysis is low. The reduced cytochrome *c* concentration would be low as the equilibrium of the reactions shift back to formation of NADH and FADH$_2$, and oxidative phosphorylation is minimal. Consequently, when the cell has plenty of ATP, catabolic pathways are generally inhibited, and the cell can do more synthetic pathways.

COUPLED VERSUS UNCOUPLED MITOCHONDRIA

As has been discussed, the movement of electrons through Complexes I through IV produces a proton gradient, which is used by Complex V for ATP synthesis. As mentioned in the regulation of the electron transport chain, if ATP synthesis stops, so does oxygen consumption, electron movement, and proton pumping. This is because the activities of Complexes I through IV are *coupled* to Complex V, the ATP synthase.

In a coupled mitochondria, when Complex V is inhibited because there is plenty of ATP, electron movement via Complexes I through IV also stops because the proton gradient is not being used. Envision that there are only so many protons that can be put in the intermembrane space of a mitochondrion. Unless there is some means of releasing the protons from the intermembrane space, electron movement and oxygen consumption will also stop.

As an analogy, think about a refrigerator. One stocks the refrigerator to capacity after a trip to the grocery store. Until some of the food is consumed, no more food can be put in the refrigerator, as there is a limit to how much food the refrigerator can hold. There are two basic ways the food can be removed from the refrigerator. People can consume the food, meaning useful work is done (similar to ATP synthesis by the mitochondria). A second way to empty the refrigerator is to throw the food in the garbage, wasting it so that no useful work is done (similar to energy lost as heat). Either means will allow more food to be placed in the refrigerator because food is being removed.

In *uncoupled* mitochondria, electron transfer does occur via Complexes I through IV, oxygen is consumed, and protons are pumped into the intermembrane space to form a proton gradient. However, ATP synthesis does *not* occur. Consequently, Complexes I through IV are *uncoupled* from Complex V. The proton gradient is not used to make ATP, meaning the protons are not flowing back through Complex V to the matrix. In uncoupled mitochondria, the protons are allowed to "leak out" of the intermembrane space through other means, with the energy from the membrane potential simply lost as heat (like throwing out the food from the refrigerator in the earlier analogy).

The uncoupling of mitochondria does occur naturally in a specific type of tissue called brown adipose tissue (BAT), or brown fat. Newborns (especially in the chest region) and hibernating animals have this specialized tissue. The mitochondria in BAT have a natural uncoupler, which is a special channel that makes the inner mitochondrial membrane "leaky" to protons. This special channel is called *thermogenin* or uncoupler protein 1 (UCP1). Thermogenin is located in the inner mitochondrial membrane. It is another proton channel allowing for the flow of protons from the intermembrane space to the matrix. When the protons pass through this channel, rather than Complex V, the energy of the proton gradient is given off as heat. In the first analogy, this is like pushing the rock down the hill without the rope tied to it. Using the second analogy, this is like throwing the food from the refrigerator in the garbage rather than eating it. In either case no useful work is done, and the energy is simply released as heat. Since the proton gradient is being dissipated (used) by flowing through thermogenin, electron movement, oxygen consumption, and proton pumping can continue. For a hibernating animal and a newborn, the purpose of BAT is to keep the animal or baby warm.

This natural uncoupling of mitochondria is highly controlled. Thermogenin is inhibited by GTP. Free fatty acids overcome the inhibition by GTP and open the channel, allowing protons to flow through it back to the matrix. The free fatty acids are derived from the breakdown of triacylglycerols, which are stored in adipose tissue (i.e. BAT). The generation of heat by uncoupling mitochondria in BAT is controlled by the hormone norepinephrine through a cAMP-dependent lipase. *Lipases* are enzymes that cleave lipids. Norepinephrine leads to activation of the lipases to cleave fatty acids from triacylglycerols. Heat production in adults is not controlled by brown adipose tissue, as adults have very little BAT. There are other uncoupling proteins (UCP2–UCP5) in the mitochondria of other tissues, though their specific purposes are still unknown. There are also mitochondrial poisons that act as uncouplers. One example is 2,4-dinitrophenol (2,4-DNP), which makes the mitochondrial membranes leaky to protons. This process is uncontrolled and can be fatal.

IMAGE CREDIT

- Fig. 4.3: Tiger66, "Schematic Representation of the ETC," https://commons.wikimedia.org/w/index.php?curid=20449153. Copyright in the Public Domain.

CHAPTER 5

GLUCONEOGENESIS AND THE PENTOSE PHOSPHATE PATHWAY

OBJECTIVES

1. Define the process of gluconeogenesis.
2. Explain the purpose of gluconeogenesis.
3. Identify where gluconeogenesis takes place in the cell and which tissues/organs can release free glucose into the blood.
4. Explain how gluconeogenesis is carried out in the cell.
   a. Identify the sources of pyruvate for gluconeogenesis.
   b. Identify the three irreversible reactions of glycolysis, which must be bypassed in gluconeogenesis.
   c. Describe all of the necessary reactions and transports needed in gluconeogenesis to bypass the three irreversible reactions of glycolysis, and know where in the cell these steps occur.
5. Explain when gluconeogenesis takes place.
   a. Identify the regulatory steps in gluconeogenesis, the enzymes that catalyze these steps, and the molecules (compare/contrast with glycolysis) that regulate them.
   b. Explain how reciprocal regulation of glycolysis and gluconeogenesis is accomplished and why it is important.
   c. Identify the physiological circumstances under which gluconeogenesis would be stimulated and explain why it is stimulated under these conditions.
6. Define the process of the pentose phosphate pathway.
7. Explain the purpose of the pentose phosphate pathway.
   a. Identify the uses of NADPH, ribose 5-phosphate, fructose 6-phosphate, and glyceraldehyde 3-phosphate.
8. Identify where the pentose phosphate pathway takes place in the cell.
9. Explain how the pentose phosphate pathway is carried out in the cell.
   a. Identify the substrates and products of the reactions catalyzed by glucose 6-phosphate dehydrogenase and 6-phosphogluconate dehydrogenase in the oxidative phase.
   b. Explain how the pentose phosphate pathway can generate ribose 5-phosphate and intermediates of the glycolytic pathway, and the enzymes used in the non-oxidative phase.
10. Explain when the pentose phosphate pathway takes place.
    a. Diagram how the reactions of glycolysis/gluconeogenesis and the pentose phosphate pathway can be combined to satisfy the need(s) of a cell: (1) both NADPH and ribose 5-phosphate, (2) NADPH only, (3) NADPH and ATP, and (4) ribose 5-phosphate only.
11. Explain how certain vitamin deficiencies would affect the pentose phosphate pathway.
    a. Identify transketolase as a thiamine-dependent enzyme and describe the clinical consequences of thiamine deficiency (i.e. modes can/cannot be carried out).

THREE KEY BRANCH POINTS IN METABOLISM

In this chapter two additional pathways of carbohydrate metabolism will be discussed, gluconeogenesis and the pentose phosphate pathway. In understanding how metabolic pathways are integrated (i.e., answering the question *when*), one must begin to identify key metabolites that are used in more than one pathway. These serve as "branch points" of metabolism, where regulation of pathways is important in directing these metabolites to pathways that are necessary at that time. **Figure 5.1** indicates three metabolites that are key branch points in metabolism: glucose 6-phosphate, pyruvate, and acetyl CoA. One should be able to recognize (and/or draw) them. One should also know the enzymes of the pathways that produce or use these metabolites.

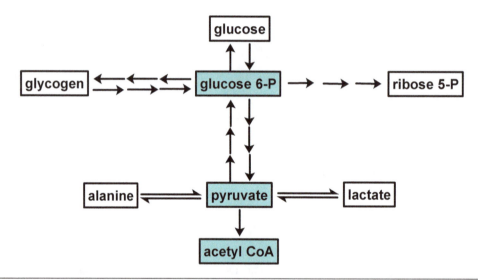

Figure 5.1: Three key branch points in metabolism are glucose 6-phosphate, pyruvate, and acetyl CoA.

Glucose 6-phosphate sits at an important branch point in metabolism. As previously discussed, glucose (upon entry into a cell from the blood) must be converted to glucose 6-phosphate before proceeding to pyruvate (or lactate) in glycolysis. Gluconeogenesis requires a special enzyme, glucose 6-phosphatase, to convert glucose 6-phosphate to glucose, as will be discussed in this chapter. Glycogen serves as the storage form of glucose. While the synthesis of glycogen (glycogenesis) and glycogen breakdown (glycogenolysis) will not be discussed in this textbook, note that glucose 6-phosphate is important to both of these pathways. Glucose 6-phosphate is the starting material for glycogen synthesis and is the product of glycogen breakdown. When glucose 6-phosphate is derived from glycogen breakdown, the fate of the glucose 6-phosphate will depend on the cell type and the needs of the cell. Only the liver and kidneys can release glucose into the blood, as will be discussed. Another possible fate of glucose 6-phosphate is entry into the pentose phosphate pathway.

Pyruvate is another important branch point. The fate of pyruvate depends on the energy state of the cell and if oxygen is available. If oxygen is not available, lactate may be formed. If oxygen is available, pyruvate may be transported into the mitochondria for formation of acetyl CoA by the PDH complex. If alanine is needed, pyruvate may be converted to alanine via a transamination reaction. Lastly, the pyruvate formed may be coming from amino acid catabolism (such as alanine) or from lactate for entry into gluconeogenesis. Acetyl CoA is a third important branch point of metabolism. As will be discussed in the lipid chapters, acetyl CoA is the major product of fatty acid catabolism. The acetyl CoA derived from the breakdown of fatty acids can enter the TCA cycle or, in the liver only, be used for ketone body synthesis. The starting material for fatty acid synthesis is also derived from acetyl CoA, and the catabolism of several of the amino acids also yields acetyl CoA.

The importance of these intermediates as branch points between several metabolic pathways can be demonstrated by envisioning what happens when a person has an enzyme defect in a particular pathway. As an analogy, if there is an accident on the interstate, there is little to no traffic in front of the accident, and traffic backs up behind it. An interstate is called a limited access highway, since there are only certain points that cars can get on and off. When traffic backs up due to an accident, it will back up to an exit, where one will then see increased traffic leaving the highway. If the accident is not cleared up quickly, traffic will continue to back up even further to another exit ramp (or branch point). The same thing happens in metabolism. If there is an enzyme defect, or something is preventing an enzyme or reaction from being carried out, the pathway cannot continue, and the prior intermediates of that pathway will build up until a branch point is encountered. These intermediates will then be "shuttled through" alternate pathways to alleviate the buildup of the intermediates in the blocked pathway. In a patient who has an enzymatic defect, determination of metabolites that are in excess versus deficient can help pinpoint where the defect is.

GLUCONEOGENESIS

Gluconeogenesis (*what*) is a metabolic pathway that synthesizes glucose from pyruvate. Pyruvate, as the starting material, is obtained from non-carbohydrate carbon sources, such as lactate, glucogenic amino acids, and glycerol. Humans cannot use products from the catabolism of fatty acids or ketogenic amino acid products for the synthesis of glucose. The catabolism of fatty acids and ketogenic amino acids primarily yields acetyl CoA. The pyruvate dehydrogenase (PDH) complex is irreversible, as previously covered in chapter 3. Consequently, acetyl CoA cannot be directly converted to pyruvate. Also, as the two-carbon acetyl CoA enters the TCA at the formation of citrate, two carbons are lost as CO_2 before reaching oxaloacetate, which is also an intermediate of gluconeogenesis. Thus, humans cannot achieve net synthesis of glucose from acetyl CoA from the catabolism of fatty acids or certain amino acids.

Glucose serves as the key circulating fuel source for the body. Glucose is the primary fuel source for the brain. Also, mature red blood cells can only use glucose as a fuel source, since they lack mitochondria. While the catabolism of a fatty acid provides more ATP per molecule than a single glucose molecule, the hydrophobic nature of fatty acids makes them difficult to keep in high concentration in the blood.

Therefore, maintaining a constant circulating concentration of glucose in the blood is a metabolic priority. The purpose of gluconeogenesis (*why*) is to make glucose for storage purposes, as glycogen, or glucose can be released directly into the bloodstream. The liver and kidneys, though, are the only organs that can release glucose (sometimes called "free glucose") into the blood. The liver carries out approximately 90 percent of the gluconeogenic activity of the body. The kidneys serve to back up this function of the liver by carrying out about 10 percent of the gluconeogenic activity in the body. The glucose supplied by the liver and kidneys is taken up by the other tissues, notably the brain, heart, muscle, and red blood cells. Lactate and amino acids released by these tissues into the bloodstream can then be taken up by the liver and kidneys to supply pyruvate for gluconeogenesis.

A typical adult liver can store enough glucose, as glycogen, to meet the needs of the body for about twenty-four hours. Infants and children have much less glucose reserves in their livers, from about a two-hour supply in infants to about a four- to six-hour supply in young children (less than ten years old). Consequently, infants and children are more dependent on the pathway of gluconeogenesis to maintain their blood glucose levels between meals and overnight than an adult. As a clinical note, the metabolism of alcohol inhibits gluconeogenesis. Inhibition of gluconeogenesis in young children by ingestion of alcohol can quickly induce severe hypoglycemia.

While the pathways of gluconeogenesis and glycolysis share many of the same reactions, gluconeogenesis is not an exact reversal of glycolysis. Glycolysis is an overall exergonic pathway, allowing for the synthesis of ATP. Gluconeogenesis, on the other hand, is a synthetic pathway in which smaller starting materials (two pyruvate) are ultimately converted to molecules that can be covalently bound to make a larger product, glucose. This overall process is endergonic and requires the input of energy, as shown in **figure 5.2**. Glycolysis has three irreversible enzymatic reactions, as discussed in chapter 2. Gluconeogenesis will have to bypass these reactions in some manner to achieve the synthesis of glucose.

$$\text{2 pyruvate + 4 ATP + 2 GTP + 2 NADH + 6 H}_2\text{O} \longrightarrow \text{glucose + 4 ADP + 2 GDP + 6 P}_i\text{ + 2 NAD}^+\text{ + 2 H}^+$$

The free energy change for gluconeogenesis (above) is: $\Delta G^{0\prime} = -9$ kcal/mol
The exact reversal of glycolysis, however, is: $\Delta G^{0\prime} = +20$ kcal/mol

Figure 5.2: Net reaction of gluconeogenesis.

Another distinction between glycolysis and gluconeogenesis is where the pathway takes place in the cell. Reciprocal pathways cannot be active, or on, in the same cell at the same time, as that would produce a futile cycle. Many synthetic pathways occur in the cytosol, while their reciprocal catabolic pathways occur in the mitochondria. However, glycolysis is an important catabolic pathway that must reside outside of the mitochondria in the cytosol, as discussed in chapter 2. As both pathways share many of the same reactions, another level of regulation of these reciprocal pathways is to compartmentalize part of gluconeogenesis in the mitochondria. The first step(s) of gluconeogenesis take place (*where*) in the mitochondrial matrix, while the remaining steps take place in the cytosol. Most of the gluconeogenic reactions that take place in the cytosol use the same glycolytic enzymes, only going in reverse. As mentioned, the liver and kidneys are the primary tissues that carry out gluconeogenesis.

The complete pathway of gluconeogenesis is shown in **figure 5.3**, as well as how it compares to glycolysis. As mentioned, gluconeogenesis is not an exact reversal of glycolysis, which is essentially due to the three irreversible steps of glycolysis. The three irreversible steps of glycolysis are catalyzed by hexokinase or glucokinase (**figure 2.5**), phosphofructokinase (**figure 2.7**), and pyruvate kinase (**figure 2.15**). These three irreversible steps of glycolysis are bypassed by *four* new reactions unique to gluconeogenesis. The remaining steps of gluconeogenesis are simply the reverse reactions of glycolysis and use the same enzymes that were

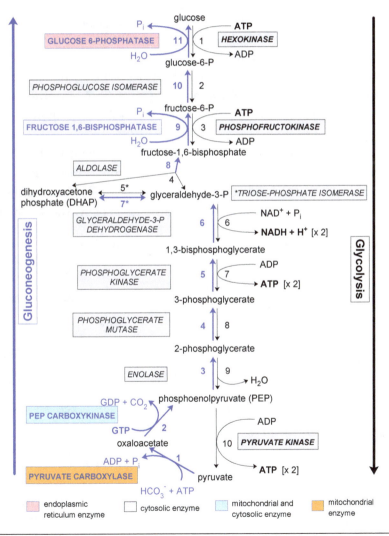

Figure 5.3: The reciprocal pathways of gluconeogenesis and glycolysis.

described in detail in chapter 2. Only the four unique reactions of gluconeogenesis will be covered in further detail for this pathway (*how*).

The first part of gluconeogenesis occurs in the mitochondrial matrix. The generation of pyruvate from non-carbohydrate precursors, such as lactate and alanine, generally occurs in the cytosol. The cytosolic pyruvate must cross the inner mitochondrial membrane via a pyruvate transporter, which is the same transporter needed to import pyruvate for the PDH complex.

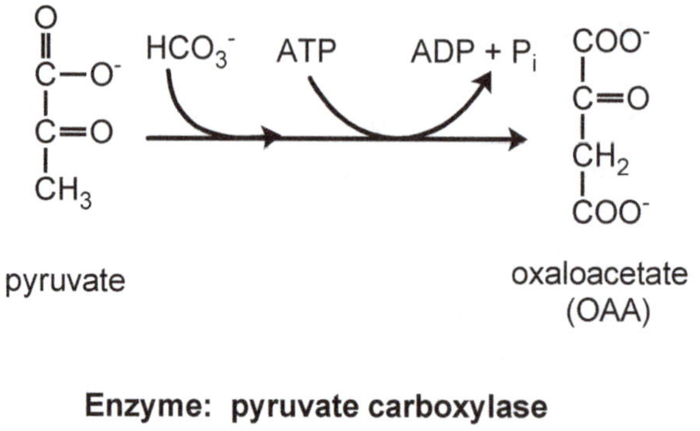

Enzyme: pyruvate carboxylase

Figure 5.4: Reaction 1: Carboxylation of pyruvate to oxaloacetate.

The first reaction of gluconeogenesis is the carboxylation of pyruvate (a three-carbon molecule) to oxaloacetate (a four-carbon molecule), as shown in **figure 5.4**. Pyruvate and oxaloacetate are molecules shown previously and are key structures one should know. Enzymes that carry out carboxylation reactions are called *carboxylases* and are named for the molecule accepting the CO_2 group. Thus, the enzyme name for this reaction is *pyruvate carboxylase*. Pyruvate carboxylase is a mitochondrial enzyme.

Pyruvate carboxylase requires bicarbonate (as the source of CO_2) and ATP. Carboxylase enzymes also require the coenzyme biotin. **Figure 5.5** shows the two steps of the pyruvate carboxylase enzyme. Pyruvate carboxylase first picks up the CO_2 group and attaches it to the biotin coenzyme, which requires the use ATP. The carboxylic acid group is then transferred to the methyl carbon of pyruvate, forming oxaloacetate.

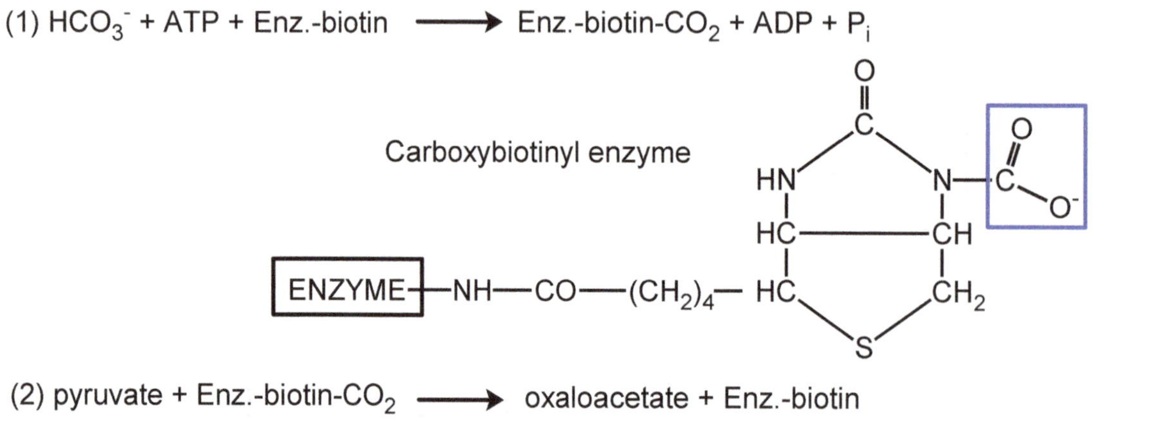

(1) HCO_3^- + ATP + Enz.-biotin $\longrightarrow$ Enz.-biotin-CO_2 + ADP + P_i

Carboxybiotinyl enzyme

(2) pyruvate + Enz.-biotin-CO_2 $\longrightarrow$ oxaloacetate + Enz.-biotin

Figure 5.5: The two steps of the pyruvate carboxylase reaction involving biotin.

The next reaction of gluconeogenesis converts oxaloacetate to phosphoenolpyruvate, shown in **figure 5.6**. In humans, *phosphoenolpyruvate carboxykinase* (abbreviated as PEP carboxykinase or PEPCK), the enzyme that catalyzes this reaction, is found in *both* the mitochondrial matrix and the cytosol. The inner

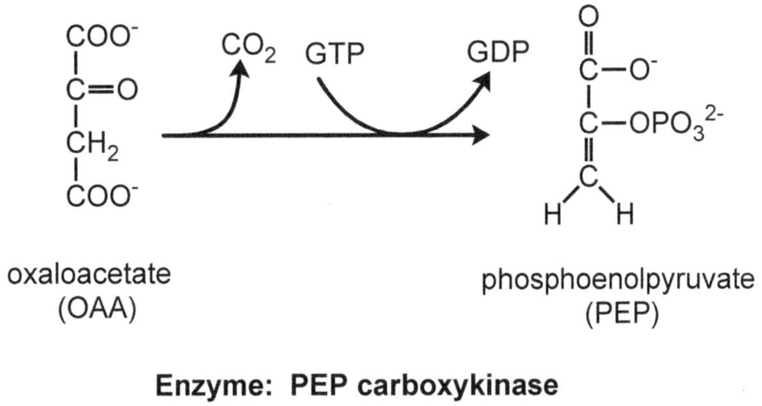

Enzyme: PEP carboxykinase

Figure 5.6: Reaction 2: Decarboxylation and phosphorylation to convert oxaloacetate to phosphoenolpyruvate.

mitochondrial membrane has a specific transporter to transport the phosphoenolpyruvate generated in the mitochondria to the cytosol for the remainder of gluconeogenesis. Recall, though, oxaloacetate cannot cross the inner mitochondrial membrane because there is no transporter for it. To move some of the oxaloacetate to the cytosol for the cytosolic reaction of converting oxaloacetate to phosphoenolpyruvate, the malate-aspartate shuttle (**figure 2.18**) must run in reverse (i.e., start with oxaloacetate in the matrix in **figure 2.18**). Oxaloacetate is first reduced to malate (which can cross the inner mitochondrial membrane) by *mitochondrial malate dehydrogenase*. The malate is then transported to the cytosol, where it is re-oxidized to oxaloacetate by *cytosolic malate dehydrogenase*.

PEP carboxykinase catalyzes both steps of the reaction needed to convert oxaloacetate to phosphoenolpyruvate (**figure 5.6**). The first step is the decarboxylation of oxaloacetate (four carbons), which drives the second step of phosphorylation to form phosphoenolpyruvate (three carbons). This reaction uses GTP as the phosphate donor, rather than ATP. While it seems wasteful to put a carboxylic group on a molecule only to remove it in the following step, the decarboxylation provides the energy needed to drive an otherwise endergonic reaction. A similar example of this use of a decarboxylation reaction is seen in fatty acid synthesis in chapter 7.

Once phosphoenolpyruvate is formed, the reactions that occur from phosphoenolpyruvate to the formation of fructose 1,6-bisphosphate are all simply the reverse reactions of glycolysis and use the same enzymes as discussed in chapter 2. The next unique reaction of gluconeogenesis is the reaction of fructose 1,6-bisphosphate to fructose 6-phosphate, shown in **figure 5.7**. Recall that alcohol phosphate groups only have a $\Delta G°'$ of about −3 kcal/mol. To reverse the corresponding glycolytic reaction, catalyzed by phosphofructokinase, would require the synthesis of ATP using the phosphate cleaved from carbon 1 of fructose 1,6-bisphosphate. As there is not enough energy in this bond, this new enzyme of gluconeogenesis simply hydrolyzes (i.e., adds water across the bond) the phosphate group from carbon 1, forming P_i as a product, along with fructose 6-phosphate.

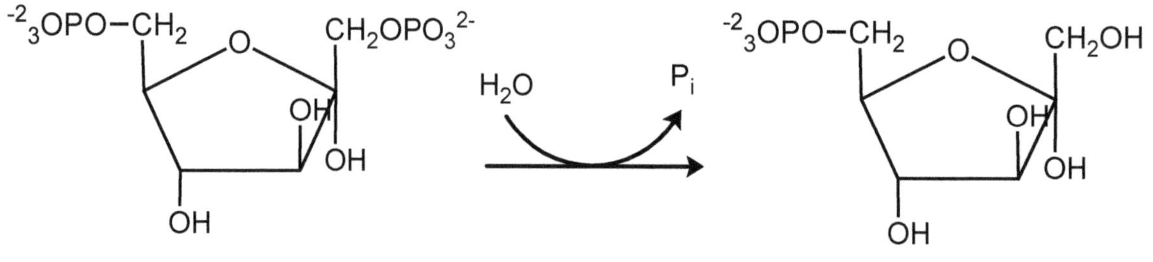

Enzyme: fructose 1,6-bisphosphatase (FBP1)

Figure 5.7: Reaction 9: Dephosphorylation of fructose 1,6-bisphosphate to fructose 6-phosphate.

Enzymes that remove phosphate groups by hydrolysis (i.e., dephosphorylate) are called *phosphatases* and are named for the molecule from which the phosphate group is removed. Hence, the enzyme for this reaction is called *fructose 1,6-bisphosphatase* (FBP1). FBP1 catalyzes an *irreversible* reaction of gluconeogenesis and is highly regulated.

The last unique reaction of gluconeogenesis is the last step of the pathway, the conversion of glucose 6-phosphate to glucose (**figure 5.8**). In this reaction, glucose 6-phosphate is hydrolyzed at the phosphate ester of carbon 6 to yield glucose and P_i. Again, to reverse the corresponding reaction of glycolysis, catalyzed by hexokinase or glucokinase, would require the synthesis of ATP. As mentioned for the fructose 1,6-bisphosphatase reaction, there is not inherently enough energy in the alcohol phosphate bond to synthesize a phosphoanhydride bond. Therefore, the phosphate group is simply clipped off and released as P_i.

Enzymes that catalyze dephosphorylation reactions are called *phosphatases* and are named for the molecule from which the phosphate group is removed. Thus, the enzyme for this reaction is called *glucose 6-phosphatase*. This step is actually carried out in the endoplasmic reticulum (not the cytosol). Glucose

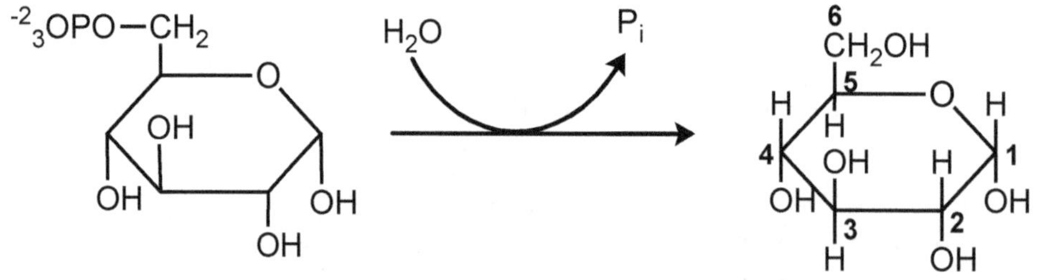

Enzyme: glucose 6-phosphatase

Figure 5.8: Reaction 11: Dephosphorylation of glucose 6-phosphate to glucose.

UNDERSTANDING BIOCHEMICAL PATHWAYS: A PATTERN-RECOGNITION APPROACH

6-phosphate is transported into the lumen of the endoplasmic reticulum, where it is hydrolyzed to form glucose. Glucose 6-phosphatase is a membrane-bound enzyme of the endoplasmic reticulum. Glucose and P_i are then transported back to the cytosol. The glucose can then be exported to the bloodstream.

Glycogen is the body's storage form of glucose. The breakdown of glycogen (glycogenolysis) also yields glucose 6-phosphate as its product. Glucose 6-phosphatase is needed to release glucose into the blood from glycogen breakdown as well. *Glucose 6-phosphatase is only found in the liver and kidneys.* Glucose 6-phosphatase is *not* present in the brain, muscle, and other tissues. Muscle tissue, in particular, has a larger glycogen store than the liver. Because muscle tissue lacks glucose 6-phosphatase, the glycogen stores of muscle tissue cannot be used to replenish blood glucose levels. Therefore, only the liver and kidneys are responsible for maintaining blood glucose levels.

Regulation of gluconeogenesis

Gluconeogenesis and glycolysis are *reciprocally* regulated. Both pathways cannot be active at the same time in the same cell. Consequently, when one pathway is active, the other pathway is inactive in a given cell. The regulation discussed for this pathway will focus on the two basic concepts of metabolic pathway regulation stressed in this text (*when*). Namely, how the energy state of the cell affects the activation or inhibition of this pathway; and when there are two enzymes in the same compartment using the same reactants or products, how the cell regulates which enzyme is active versus inactive.

Pyruvate carboxylase, the mitochondrial enzyme that catalyzes the first reaction of gluconeogenesis, illustrates both regulatory principles. Pyruvate carboxylase is inhibited by ADP. When ADP concentrations in a cell are high (relative to ATP), the cell is low on energy and cannot do synthetic pathways. The cell needs to carry out catabolic pathways to build up its ATP pool, instead.

Also, pyruvate carboxylase and the PDH complex are both located in the mitochondrial matrix. Both of these enzymes use pyruvate as a reactant. The cell typically would only need one of these enzymes active, while the other one is inhibited. The cell controls the activity of these enzymes with a single effector molecule, acetyl CoA. Recall that the PDH complex is inhibited by high concentrations of acetyl CoA, as discussed in chapter 3. Pyruvate carboxylase, on the other hand, is *activated* by acetyl CoA. In fact, the biotin of pyruvate carboxylase is not carboxylated unless acetyl CoA (or the closely related acyl CoA) is bound to the enzyme.

The catabolism of fatty acids, covered in chapter 6, provides a high concentration of acetyl CoA in the mitochondrial matrix. Recall that a primary function of the liver is to maintain blood glucose levels—and in general, circulating fuels in the form of both glucose and ketone bodies (derived from fatty acid breakdown). Under conditions where gluconeogenesis is activated in the liver, lipid breakdown also increases to supply acetyl CoA for ketone body synthesis. Acetyl CoA serves as an important allosteric effector to keep the pyruvate carboxylase active and the PDH complex inactive. Furthermore, as the oxaloacetate is moved to the cytosol (via malate) for continuation of gluconeogenesis, there is no oxaloacetate available for the acetyl CoA to combine with to form citrate (TCA cycle). The acetyl CoA is then diverted to ketone body synthesis (another pathway only carried out by the liver and discussed in chapter 6). In this way the liver carries out gluconeogenesis and ketone body synthesis simultaneously to ensure plenty of circulating fuel sources are available in the blood.

FBP1 is another key regulated enzyme. It is activated by high levels of ATP, but inhibited by AMP (cellular ATP levels are low). FBP1 is also inhibited by fructose 2,6-bisphosphate, an allosteric effector of both FBP1 and phosphofructokinase (PFK1) of glycolysis because both of these enzymes use the same reactants/products. The reciprocal regulation of FBP1 and PFK1 was covered in chapter 2 and was shown in **figure 2.16**. In summary, fructose 2,6-bisphosphate is a signal molecule derived from fructose 6-phosphate. Phosphofructokinase 2 (PFK2) makes fructose 2,6-bisphosphate, while *fructose 2,6-bisphosphatase* (FBP2) converts it back to fructose 6-phosphate. Fructose 2,6-bisphosphate, the allosteric effector, is low during times of starvation (glucagon is the predominant hormone) favoring gluconeogenesis. This allows the liver to supply glucose to the blood. Fructose 2,6-bisphosphate is high in the fed state (insulin is the predominant hormone) favoring glycolysis by activating PFK1 and inhibiting FBP1.

FBP2 is the enzyme that is hormonally controlled. Glucagon (low blood glucose indicator) activates FBP2 (thus PFK2 is off), favoring gluconeogenesis in liver cells. Insulin (high blood glucose indicator) inhibits FBP2; thus, PFK2 is now on, favoring glycolysis. When determining which pathway, glycolysis or gluconeogenesis, is active (indicating the opposing pathway is inactive), remember the following: When the *kinases are on* (PFK1 and PFK2), *glycolysis* is active in the cell. When the *bisphosphatases are on* (FBP1 and FBP2), *gluconeogenesis* is active in the cell.

THE PENTOSE PHOSPHATE PATHWAY

The pentose phosphate pathway (*what*) is a glucose-based metabolic pathway that begins with glucose 6-phosphate and can ultimately form fructose 6-phosphate and glyceraldehyde 3-phosphate to link it back to the pathways of glycolysis or gluconeogenesis. In this way one can think of the pentose phosphate pathway as a shunt off of glycolysis (or gluconeogenesis). Thus, the location of this pathway (*where*) is in the cytosol of all cells.

The purpose (*why*) of this pathway is to make NADPH, the form of reducing power used for synthetic reactions (and other uses) and various three-, four-, five-, six-, and seven-carbon sugars, including ribose 5-phosphate needed for nucleic acid (DNA and RNA) synthesis. The pentose phosphate pathway serves as a link, then, between carbohydrate and nucleotide metabolism. The links between the pentose phosphate pathway and glycolysis/gluconeogenesis allows for the synthesis and catabolism of these pentose sugars.

NADPH has many uses, some of which depend on the cell type. NADPH is the common form of reducing power used for synthetic pathways (i.e., going up the oxidation states flowchart, **figure 1.7**), as will be seen in the fatty acid synthesis pathway in chapter 7. NADPH is used by phagocytic cells to create the respiratory burst to kill microorganisms taken up by these cells. Once the phagocytic cell engulfs a bacteria (or other foreign cell), it ultimately forms a phagolysosome. Within the phagolysosome, NADPH and O_2 are used by the enzyme *NADPH oxidase* to produce superoxide anion, the first reactive oxygen species. Superoxide anion, along with other reactive oxygen species like hydrogen peroxide (H_2O_2) and hypochlorous acid (HOCl), kill the bacteria.

Another important use of NADPH is in the synthesis of nitric oxide. The enzyme nitric oxide synthase oxidizes NADPH + H^+ to $NADP^+$ in the reduction of molecular oxygen to form nitric oxide, using arginine as the nitrogen donor. Nitric oxide is important in numerous biological functions, including neurotransmitter functions, relaxation of smooth muscle, prevention of platelet aggregation, and bactericidal actions of macrophages.

Cytochrome P450 (CYP) monooxygenase systems also require NADPH. P450 refers to the absorbance at 450 nm of the protein. CYPs require both molecular oxygen and NADPH. The term "monooxygenase" refers to the incorporation of only one of the oxygen atoms onto the substrate as a hydroxyl group (–OH group). The other oxygen atom is reduced to water. The basic reaction carried out by CYP systems is:

$$R-H + O_2 + NADPH + H^+ \rightarrow R-OH + H_2O + NADP^+$$

The substrate is denoted as R–H and may be an intermediate in bile acid or steroid synthesis, a drug, or any number of other foreign compounds.

There are two CYP monooxygenase systems, the mitochondrial system and the microsomal system. The mitochondrial system is located in the liver, kidneys, adrenal cortex, ovaries, and placenta. Depending on the tissue type, this system is important in steroid hormone synthesis, bile acid synthesis, and the formation of the active form of vitamin D. The microsomal system is located in the smooth endoplasmic reticulum, primarily in the liver, and is responsible for the hydroxylation of xenobiotics (i.e., foreign compounds, such as drugs, pesticides, and pollutants), which often makes them more soluble for excretion.

NADPH is also used in the reduction of reactive oxygen species. When oxygen acquires electrons to only partially reduce it, various reactive oxygen species are formed, including superoxide anion, hydrogen peroxide, and hydroxyl radicals. Reactive oxygen species are highly oxidative (meaning they are looking for more electrons). A reactive oxygen species will react with essentially the nearest molecule to obtain electrons to further reduce it and oxidize the molecule with which it reacts, creating oxidative stress for the cell. These molecules could be enzymes (or other proteins), membrane components, or nucleotides. Therefore, cells need to remove reactive oxygen species as quickly as possible to prevent damage to the cell. NADPH is necessary to remove them.

Red blood cells are an example cell type that must maintain a constant reducing atmosphere to quickly reduce reactive oxygen species, as it transports molecular oxygen throughout the body. There are several enzymes used to convert reactive oxygen species to harmless products. These include *catalase, superoxide*

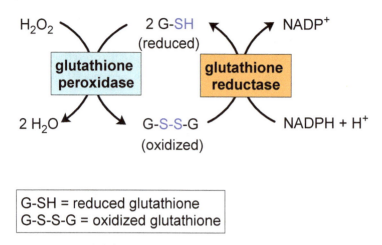

Figure 5.9: Glutathione peroxidase and glutathione reductase reactions.

dismutase, and the enzymes responsible for cycling glutathione between its oxidized and reduced forms, as shown in **figure 5.9**. Reduced glutathione (G–SH) is a tripeptide-thiol (γ-glutamylcysteinylglycine). Reduced glutathione is used to chemically reduce hydrogen peroxide (H_2O_2) to water. Cysteine has a sulfhydryl group (–SH), a.k.a. thiol group, on its side chain, which is its reduced form. *Glutathione peroxidase* (**figure 5.9**) reduces hydrogen peroxide to two water molecules by reacting two molecules of the reduced form of glutathione, forming a disulfide bond. In this reaction two hydrogen molecules are removed from the two cysteines to form the disulfide bond, which is now oxidized.

Glutathione, now in its oxidized form (abbreviated G-S-S-G), needs to be reduced back to the G-SH form to carry out further reductions of hydrogen peroxide. NADPH is used by the enzyme *glutathione reductase* (**figure 5.9**) to reduce the disulfide bond of the oxidized form of glutathione (G-S-S-G) back to two molecules of reduced glutathione (two G-SH). Note that the enzyme using the NADPH is properly called a *reductase*, and reductases are always named for the *more oxidized* molecule. However, both the substrate and product are called "glutathione," so the enzyme name is *glutathione reductase*.

Red blood cells depend on the pentose phosphate pathway for constant NADPH production to deal with the reactive oxygen species generated while transporting molecular oxygen. As a clinical note, people with a genetic defect in glucose 6-phosphate dehydrogenase (the first enzyme of the pentose phosphate pathway) are often fine and experience no symptoms until they are put under some type of additional oxidative stress. Under additional oxidative stress, their cells cannot produce enough NADPH to handle the extra reactive oxygen species that are generated. Antimalarial drugs are one example of something that can induce additional oxidative stress. People who have a genetic defect in glucose 6-phosphate dehydrogenase often never experience any problems until they have to travel to a location in which they need to take antimalarial drugs. Their mutant form of glucose 6-phosphate dehydrogenase cannot maintain the necessary levels of the reduced form of glutathione. They then experience side effects from the drug, such as anemia.

The pentose phosphate pathway consists of eight enzymatic reactions (*how*). The first three steps are called the "oxidative phase," as this is the part of the pathway that produces NADPH. The oxidative phase of the pentose phosphate pathway is also *irreversible*. The remaining five reactions are called the non-oxidative phase and constitute the reversible portion of the pathway. The non-oxidative phase produces the various other carbohydrates and provides connections to glycolysis and gluconeogenesis via fructose 6-phosphate and glyceraldehyde 3-phosphate. A key focus of this pathway is that it is a flexible pathway, which means it can do various sequences of these reactions based on the needs of the cell. These different sets of reactions will constitute the "four modes" of the pentose phosphate pathway that will be discussed further.

The eight reactions of the pentose phosphate pathway are shown in **figure 5.10** (without the structures). The intermediates that connect the pentose phosphate pathway to glycolysis and gluconeogenesis are highlighted in green. **Figure 5.11** shows the oxidative phase (irreversible phase) of the pentose phosphate pathway. Recall exercise 6 of the problem set at the end of chapter 1 on oxidation states. The answer to exercise 6 explained two different ways to get from the first molecule to the last molecule, though one of the ways was preferred. This exercise did not ask you to identify the enzymes nor the coenzymes used, because they are exceptions to the rules regarding oxidation-reduction reactions. However, exercise 6 is the reaction sequence of the oxidative phase of the pathway (minus one reaction).

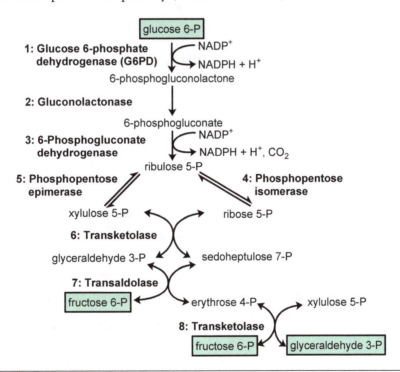

Figure 5.10: The pentose phosphate pathway.

As shown in **figure 5.11**, reaction 1 starts with glucose 6-phosphate and oxidizes the aldehyde carbon 1 to a carboxylic acid, forming 6-phosphogluconate. The name "6-phosphogluconate" tells you its structure, as long as one knows the straight chain form of glucose. Draw the straight chain form of glucose, except make

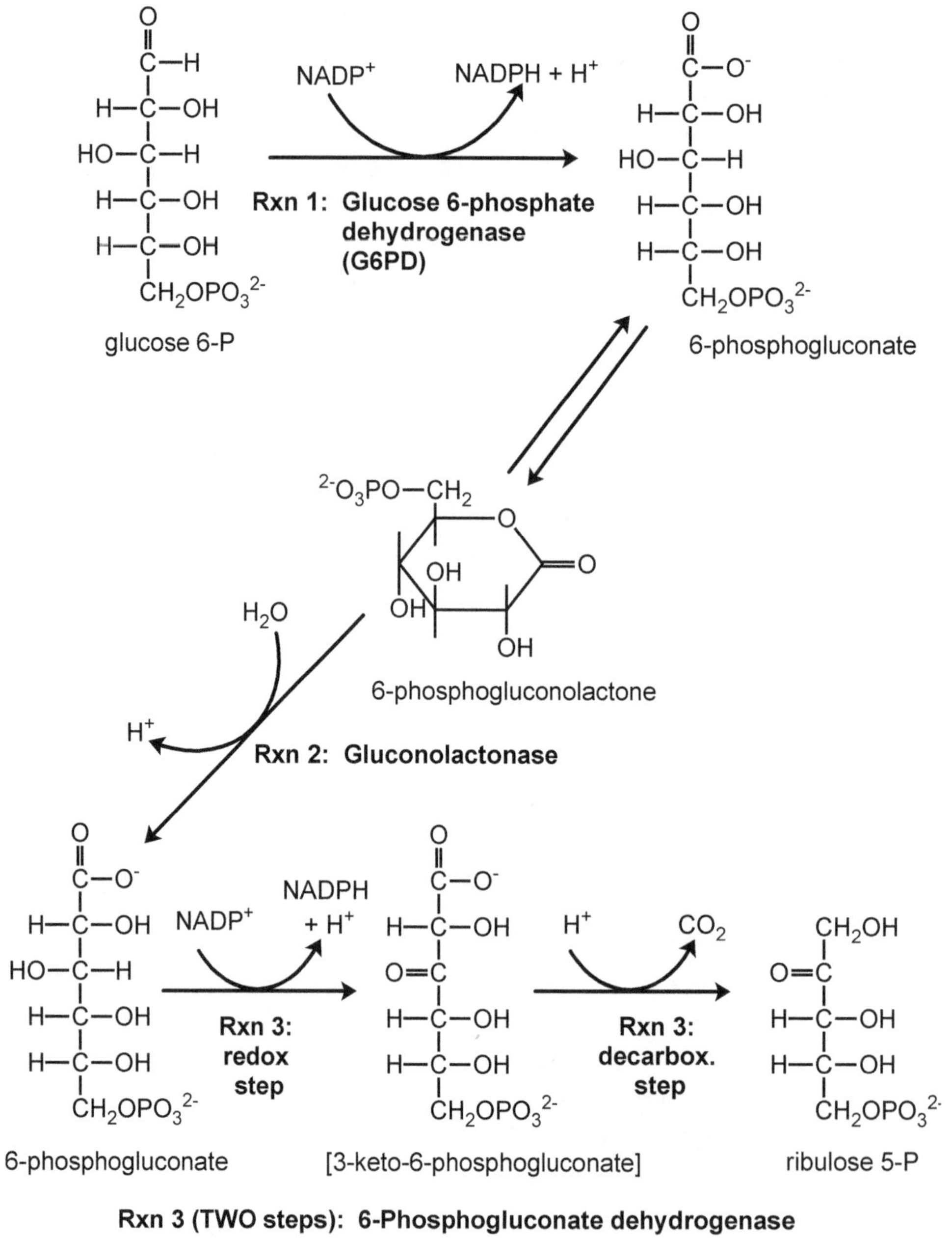

Figure 5.11: The oxidative phase (irreversible) of the pentose phosphate pathway.

carbon 1 a carboxylate ion (indicated by the "-ate" ending of the molecule name). Then draw a phosphate group on the hydroxy group of carbon 6. Typically moving down the oxidation states flowchart, the oxidation from an aldehyde to a carboxylic acid requires the reduction of NAD⁺ to NADH + H⁺. However, the purpose of the oxidative phase of the pentose phosphate pathway is to make NADPH, the reducing power needed for synthetic reactions and so on, as discussed previously. Therefore, this reaction is an *exception* to the oxidation states rule of which form of reducing power to use. This oxidation of carbon 1 to a carboxylic acid simultaneously reduces *NADP⁺* to *NADPH + H⁺*. Since the enzyme uses NADP⁺/NADPH, one would expect the enzyme to be called a reductase, named for the more oxidized molecule. However, this enzyme is also an *exception* to the naming rules. The enzyme is a *dehydrogenase*. Dehydrogenases are always named for the more reduced molecule, so the enzyme name is *glucose 6-phosphate dehydrogenase* (often abbreviated G6PD).

Glucose 6-phosphate dehydrogenase is the primary regulation point (*when*) of the pentose phosphate pathway, since the reaction it catalyzes is *irreversible*. NADPH, a product of this reaction, serves as a potent competitive inhibitor of glucose 6-phosphate dehydrogenase. The cellular ratio of NADPH/NADP⁺ determines the flux of glucose 6-phosphate through this enzyme. Under most metabolic conditions, the cellular ratio is sufficiently high to inhibit glucose 6-phosphate dehydrogenase and prevent substantial entry of glucose 6-phosphate into the pentose phosphate pathway. Metabolic conditions that increase the use of NADPH lower the ratio (i.e., higher [NADP⁺] relative to [NADPH]) and enhance the activity of this enzyme.

The straight chain form, 6-phosphogluconate, is not energetically stable. Once formed, it quickly cyclizes to the lactone form, which is more energetically stable. The formation of the lactone, 6-phosphogluconolactone, and the enzyme that cleaves it back to the straight chain form are the steps left out of exercise 6 at the end of chapter 1. Prior to carrying out the next oxidation step, the lactone must be reopened to the straight chain form of 6-phosphogluconate, as shown in reaction 2 in **figure 5.11**. A *lactonase* is an enzyme that hydrolyzes a lactone ring. Hence, the enzyme that catalyzes this reaction is called *gluconolactonase*.

Reaction 3, shown in **figure 5.11**, is a two-step reaction sequence carried out by a single enzyme. Another major product of the pentose phosphate pathway (besides NADPH) is the five-carbon sugar ribose 5-phosphate, which is needed for nucleotide synthesis. The molecule 6-phosphogluconate has a carboxylic acid group at carbon 1 that can potentially be removed. Recall the removal of carboxylic acids from a molecule requires a keto group placed either alpha (carbon 2) or beta (carbon 3) to the carboxylic acid group. The removal of carboxylic acids, though, is easier if the keto group can be placed beta to the carboxylic acid. The molecule 6-phosphogluconate has a hydroxy group on both carbon 2 and carbon 3, so either oxidation is theoretically possible (as explained by the two different answers for exercise 6 in chapter 1).

In metabolism, though, carbon 3 is oxidized to a ketone, forming a β-keto acid. This is an oxidation, so a coenzyme needs to be reduced. Normally, NAD⁺ would be reduced to NADH + H⁺. Again, a primary purpose of the oxidative portion of this pathway is to produce NADPH. This is the second reaction of the pentose phosphate pathway that is an *exception* to the coenzyme form used: *NADP⁺ is reduced to NADPH + H⁺*. This intermediate now has an unstable carboxylic acid group that can leave spontaneously. Since "spontaneously" does not equal "instantaneously," this same enzyme catalyzes the decarboxylation step forming the product ribulose 5-phosphate.

Anytime an enzyme carries out more than one reaction, if one of the reactions is an oxidation-reduction reaction, that naming takes precedence. Knowing that this enzyme uses NADP⁺/NADPH, one would expect the name to be reductase. However, this enzyme is another *exception* to the naming rules. The name of the enzyme is *dehydrogenase*, named properly for the *more reduced* molecule. Thus, the enzyme name is

6-phosphogluconate dehydrogenase. This two-step enzymatic sequence of oxidation followed by decarboxylation is similar to the isocitrate dehydrogenase reactions of the TCA cycle.

The oxidative phase of the pentose phosphate pathway is now complete. The oxidative phase produced two molecules of NADPH and created a five-carbon sugar. Ribulose 5-phosphate, though, is not the sugar needed for nucleotide synthesis. The oxidative phase is *irreversible* because the redox reactions carried out by glucose 6-phosphate dehydrogenase and 6-phosphogluconate dehydrogenase are irreversible.

The non-oxidative phase (reversible phase) is shown in **figure 5.12**. This phase may look complicated, but there is a pattern to the types of reactions that are being carried out. The reactions that make up the non-oxidative phase are all reversible reactions, which is an important point in understanding the various modes, or set of reactions, that can be carried out by the pentose phosphate pathway. One must understand

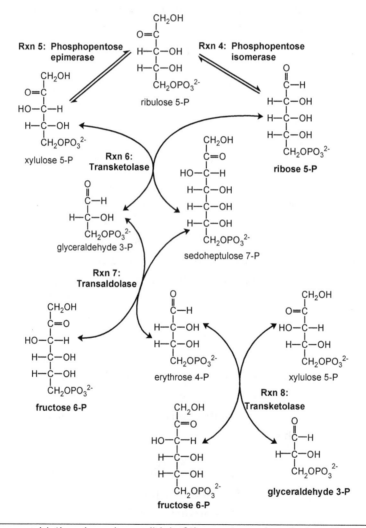

Figure 5.12: The non-oxidative phase (reversible) of the pentose phosphate pathway.

that there are pools (i.e., various concentrations) of all the reaction intermediates available, as is typical of any metabolic pathway.

To continue through the remaining five reactions, a pool of ribulose 5-phosphate must be available. Some ribulose 5-phosphate molecules will be converted to ribose 5-phosphate, while some ribulose 5-phosphate molecules will be converted to xylulose 5-phosphate. As shown in reaction 4 of **figure 5.12**, ribulose 5-phosphate is a ketose sugar, while ribose 5-phosphate (needed for nucleotide synthesis) is an aldehyde sugar. These two molecules are isomers of one another (i.e., same number of each atom type, just arranged differently). Recall from chapter 1 that ketones and aldehydes can be readily isomerized, and enzymes that carry out these types of reactions are *isomerases*. As both molecules are phosphopentoses (five-carbon phosphorylated sugars), the enzyme name is *phosphopentose isomerase*. This naming convention is similar to the enzyme name of triose phosphate isomerase in glycolysis.

Depending on the mode of the pentose phosphate pathway, some ribulose 5-phosphate may also be converted to xylulose 5-phosphate, as shown in reaction 5 of **figure 5.12**. Ribulose 5-phosphate and xylulose 5-phosphate are both ketoses and are also five-carbon phosphorylated sugars. Ribulose 5-phosphate and xylulose 5-phosphate, though, are a special type of isomer called an epimer. Epimers are molecules that are isomers of one another that differ only in the configuration at one carbon. In this case the hydroxy group on carbon 3 of ribulose 5-phosphate points to the right, while the hydroxy group on carbon 3 points to left for xylulose 5-phosphate. Enzymes that are capable of switching the configuration about a single carbon are called *epimerases*. The enzyme that interconverts these two five-carbon phosphorylated ketoses is called *phosphopentose epimerase*.

Once pools of ribose 5-phosphate and xylulose 5-phosphate are available, the pentose phosphate pathway can proceed through the remaining three reactions (reactions 6, 7, and 8). While the molecules in these three reactions may look complicated, they are all ketoses or aldoses phosphorylated at their highest-numbered carbon (i.e., the phosphate is on carbon 3, 4, 5, 6, or 7). In learning these structures, one should be able to recognize and draw the key structures of glyceraldehyde 3-phosphate, fructose 6-phosphate, and ribose 5-phosphate (based on knowing the structures of glycerol, fructose, and ribose). The relationship of ribulose 5-phosphate relative to ribose 5-phosphate and xylulose 5-phosphate has already been discussed. It should not be necessary to memorize these structures in their entirety; rather, simply note the changes that need to be made to ribose 5-phosphate to recognize and/or draw these structures. In these last three reactions, two additional sugar molecules will be formed, but the remaining sugar intermediates have already been introduced. There is a pattern to these last three reactions. Once the pattern is recognized, the reaction sequence appears more logical, and the sugar molecules should be more readily recognized and/or drawn.

For reactions 6, 7, and 8 the *donor* molecule will always be a *ketose* and the *acceptor* molecule will always be an *aldose*. Once the ketone part of the sugar is removed from the donor molecule, it is now an aldose. Once the aldose accepts the ketone portion, it becomes a ketose. The enzyme *transketolase* will be used in *both* reactions 6 and 8. Reaction 7 uses an enzyme called *transaldolase*. The simplest way to remember the difference between the transketolase versus the transaldolase is that the *trans<u>keto</u>lase cleaves at the <u>keto</u>*

group carbon. The *transketolase* removes the *top two carbons* by cleaving just after the *keto group carbon 2* (i.e., cleaves the molecule between carbons 2 and 3). The *transaldolase* removes the *top three carbons*, which does include the keto group carbon, but it cleaves between carbons 3 and 4 (i.e., "past" the carbon with the ketone group).

In reaction 6 (**figure 5.13**), *transketolase* is used to remove the top two carbons from xylulose 5-phosphate and attach them at carbon 1 of ribose 5-phosphate. The removal of the top two carbons from xylulose 5-phosphate produces the three-carbon product of glyceraldehyde 3-phosphate (a structure that has already been shown in several metabolic pathways). The placement of this two-carbon keto molecule onto carbon 1 of ribose 5-phosphate produces the seven-carbon ketose, sedoheptulose 7-phosphate. To draw this molecule, draw a seven-carbon straight chain with —CH_2OH at carbon 1, the keto group on carbon 2, and —$CH_2OPO_3^{2-}$ at carbon 7 (which is the same molecular arrangement for carbons 1, 2, and the last carbon for any of the ketoses in this pathway). The keto groups in **figure 5.12** are pointing left or right in the diagram because the keto carbon is not a chiral carbon, so one can draw it in either direction. For

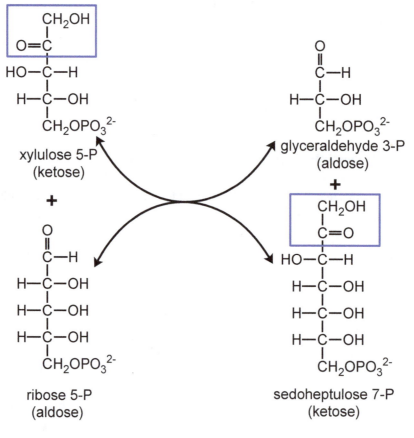

Figure 5.13: Reaction 6: Remove two carbons from xylulose 5-phosphate (ketose) and put on ribose 5-phosphate (aldose).

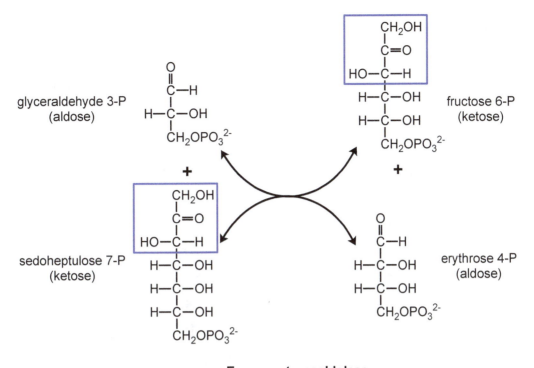

Figure 5.14: Reaction 7: Remove three carbons from sedoheptulose 7-phosphate (ketose) and put on glyceraldehyde 3-phosphate (aldose).

sedoheptulose 7-phosphate, the hydroxy groups on carbons 4, 5, and 6 all point to right (just like in ribose). Lastly, the hydroxy group on carbon 3 (formed from the aldehyde carbon of ribose 5-phosphate) points to the left—the only hydroxy group on this molecule that points to the left.

In reaction 7 (**figure 5.14**), the enzyme used is *transaldolase*, and the products from reaction 6 are the substrates of reaction 7. Transaldolase removes the top three carbons from sedoheptulose 7-phosphate (the ketose donor) by cleaving the molecule between carbons 3 and 4. The remaining four-carbon molecule is now erythrose 4-phosphate (an aldose). In drawing erythrose 4-phosphate, draw the bottom three carbons (carbons 2, 3, and 4) and their functional groups exactly as drawn in sedoheptulose 7-phosphate. Now draw the new carbon 1 (previously carbon 4 in sedoheptulose 7-phosphate) as an aldehyde. To review, erythrose 4-phosphate has an aldehyde carbon as carbon 1, the $-CH_2OPO_3^{2-}$ at carbon 4, while the two hydroxy groups on carbons 2 and 3 point to the right (compare to carbons 3, 4, and 5 of ribose 5-phosphate).

In reaction 7 (**figure 5.14**), the three-carbon molecule (including the keto group) cleaved from sedoheptulose 5-phosphate will now be placed on glyceraldehyde 3-phosphate (the aldose acceptor). Carbon 3 of the sedoheptulose 7-phosphate will be attached to the aldehyde carbon (carbon 1) of glyceraldehyde 3-phosphate. The new ketose sugar formed is fructose 6-phosphate, a structure with which one should already be familiar. If one recalls the linear structure of fructose, one can simply draw the product. If not, the top three carbons from sedoheptulose 7-phosphate remain in the same orientation, as do carbons 2 and 3 from glyceraldehyde 3-phosphate. The orientation of the new hydroxy group formed from the aldehyde carbon of glyceraldehyde 3-phosphate is the only one in question—and it will point to the right. Recall that in the linear structure of fructose, only the hydroxy group on carbon 3 points to the left. Reaction 7 makes a useful glycolytic intermediate, *fructose 6-phosphate*, from the 7-carbon sedoheptulose 7-phosphate.

xylulose 5-P (ketose)

$$\begin{array}{c} CH_2OH \\ | \\ O=C \\ | \\ HO-C-H \\ | \\ H-C-OH \\ | \\ CH_2OPO_3^{2-} \end{array}$$

glyceraldehyde 3-P (aldose)

$$\begin{array}{c} O \\ \parallel \\ C-H \\ | \\ H-C-OH \\ | \\ CH_2OPO_3^{2-} \end{array}$$

erythrose 4-P (aldose)

$$\begin{array}{c} O \\ \parallel \\ C-H \\ | \\ H-C-OH \\ | \\ H-C-OH \\ | \\ CH_2OPO_3^{2-} \end{array}$$

fructose 6-P (ketose)

$$\begin{array}{c} CH_2OH \\ | \\ C=O \\ | \\ HO-C-H \\ | \\ H-C-OH \\ | \\ H-C-OH \\ | \\ CH_2OPO_3^{2-} \end{array}$$

Enzyme: transketolase

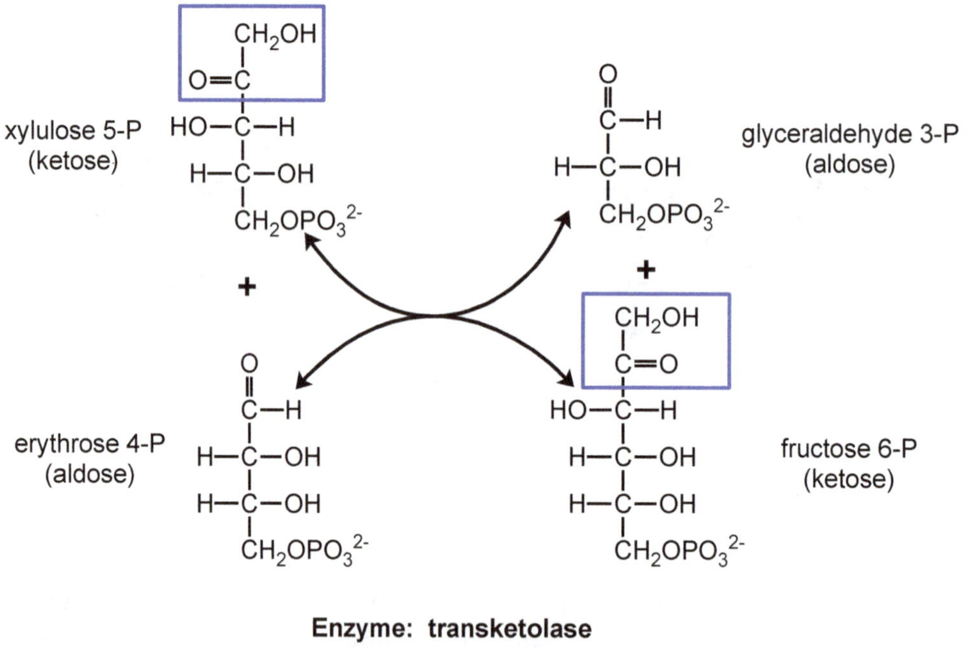

Figure 5.15: Reaction 8: Remove two carbons from xylulose 5-phosphate (ketose) and put on erythrose 4-phosphate (aldose).

In reaction 8 (**figure 5.15**), transketolase is used again. In this reaction, the transketolase removes two carbons (carbons 1 and 2) from xylulose 5-phosphate (the product of the epimerase reaction), which serves as the ketose donor. Once these two carbons are removed, the resulting three-carbon molecule is *glyceraldehyde 3-phosphate*. The two-carbon piece removed from xylulose 5-phosphate is put on a molecule of erythrose 4-phosphate (the aldose acceptor) at carbon 1. The new ketose product formed is another molecule of *fructose 6-phosphate*.

The transketolase used in reactions 6 and 8 requires *thiamine pyrophosphate (TPP)*. Recall that TPP is derived from the vitamin thiamine (B_1) and is also needed for the PDH complex and the α-ketoglutarate dehydrogenase complex. A deficiency of thiamine will thus inhibit transketolase, as well as the two complexes.

In summary, reactions 6, 7, and 8 are simply shuffling the carbons around to make different ketoses and aldoses, which may be necessary in various cellular structures or can ultimately be put back into glycolysis or gluconeogenesis. Again, the ketose is always the donor molecule, and the aldose is always the acceptor molecule for these three reactions. Transketolase always moves two carbons from a ketose to an aldose, while transaldolase always moves three carbons from a ketose to an aldose. The products of both transketolase and transaldolase reactions are new ketoses and new aldoses. As discussed, none of the actual reaction mechanisms were covered for the enzymes of the pentose phosphate pathway. The emphasis here is on recognizing the patterns of the reactions, recognizing key structures (from which other intermediates can be drawn), and how enzymes are named.

At this point, the two molecules of fructose 6-phosphate (from reactions 7 and 8) and the glyceraldehyde 3-phosphate from reaction 8 can now feed back into glycolysis (or gluconeogenesis). Depending on the needs of the cell, fructose 6-phosphate and glyceraldehyde 3-phosphate can continue through the remainder of glycolysis to yield energy. Or they may proceed through gluconeogenesis to regenerate glucose 6-phosphate. Because the non-oxidative phase of the pentose phosphate pathway is reversible, another option is for glucose 6-phosphate to first proceed through part of glycolysis to make pools of fructose 6-phosphate and glyceraldehyde 3-phosphate. Fructose 6-phosphate and glyceraldehyde 3-phosphate can then enter the pentose phosphate pathway at reaction 8 and proceed in reverse to form ribose 5-phosphate.

As previously mentioned, an important feature of the pentose phosphate pathway is that it is a very flexible pathway. What does this mean? To illustrate this concept, think about people's travel between their home and their place of work. People typically have more than one route they can travel to and from work and home. The route a person might choose on any given day may depend on traffic or errands the person may need to do en route, either going to or coming from work (i.e., stop for coffee, get gas for the car, pick up the kids from school, groceries for dinner, etc.).

The pentose phosphate pathway consists of eight reactions that produce NADPH, ribose 5-phosphate, and other carbohydrates. However, which reactions the pentose phosphate pathway carries out, and the order in which the pathway runs (forward or in reverse), depends on which of these products are needed by the cell. The cell may need to run only part of the pathway (either forward or in reverse). The cell may, instead, need to do all 8 reactions and connect back with glycolysis, if ATP is also needed; or gluconeogenesis, if the cell just needs NADPH.

Consequently, there are four different modes, or sets of reactions, that can be carried out by the pentose phosphate pathway, along with enzymes of glycolysis and gluconeogenesis. Glucose 6-phosphate is a key branch point intermediate that can go to several different pathways. How glucose 6-phosphate is utilized depends on the cell's need for ATP, NADPH, ribose 5-phosphate, or replenishing its glycogen stores. The utilization of glucose 6-phosphate by either glycolysis or the pentose phosphate pathway depends on the regulation of PFK1 and glucose 6-phosphate dehydrogenase. Both of these enzymes are highly regulated. If the pentose phosphate pathway is needed, the relative needs of ribose 5-phosphate and NADPH dictate which reactions of the pathway need to be carried out.

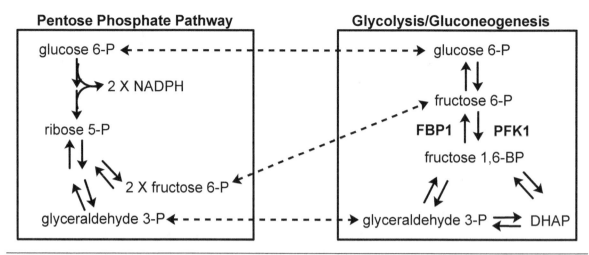

Figure 5.16: Integration of the pentose phosphate pathway with glycolysis and gluconeogenesis.

The diagram shown in **figure 5.16** provides an overview of how glycolysis and gluconeogenic reactions are connected to the pentose phosphate pathway. In this diagram the pentose phosphate pathway is outlined on the left, and part of glycolysis and gluconeogenesis are shown on the right. Glucose 6-phosphate, fructose 6-phosphate and glyceraldehyde 3-phosphate are clearly indicated as intermediates in both sets of pathways on the left and right of the diagram. The needs of the cell will then determine which set of reactions are actually carried out between glycolysis or gluconeogenesis and the pentose phosphate pathway.

The four modes will now be defined, and an example cell type or situation will be used to illustrate the mode. In *mode 1* the cell needs both ribose 5-phosphate and NADPH. An example would be cells in a growing tissue. These cells need plenty of ribose 5-phosphate for nucleic acid synthesis and NADPH for reductive biosynthesis of many cellular macromolecules (i.e., going up the oxidation states flowchart shown in **figure 1.7**). Most synthetic pathways, as will be seen for fatty acid synthesis, need both ATP and NADPH. While this is a simplistic picture, the cell will need to balance building up ATP pools and providing the ribose 5-phosphate and NADPH that are also needed by cells of a growing tissue. For illustration purposes though, in mode 1 the ATP pools are plentiful, and the cell now needs to build up its supply of ribose 5-phosphate and NADPH. In *mode 2* the cell only needs NADPH. Cells of adipose tissue, for example, generally need a lot of NADPH for fatty acid synthesis, but not necessarily ribose 5-phosphate. This situation also assumes the ATP pools are already plentiful to carry out synthetic reactions; otherwise, the cell might need to carry out mode 3.

For *mode 3* the cell needs both NADPH and ATP, but ribose 5-phosphate is not needed. A red blood cell provides a good example of this mode. Red blood cells need a constant supply of NADPH for maintaining the reduced atmosphere, as previously discussed. As mature red blood cells lack mitochondria, glycolysis is the only pathway that can provide ATP for the cell. A mature red blood cell also lacks a nucleus and therefore has no use for ribose 5-phosphate. For *mode 4* the cell only needs ribose 5-phosphate. A very rapidly proliferating cell, like a tumor cell, would need a lot of ribose 5-phosphate, but maybe not as much NADPH, if the cellular NADPH concentrations are sufficient.

Note that for each of these four modes, a theoretical cellular situation is provided. However, these situations serve as an illustration such that one may think through each possible set of reactions carried out by the cell for the given mode. In reality, most cells may carry out any one of these modes of the pentose phosphate pathway, including using some of the other sugars synthesized for other pathways.

Each of these four modes is set up as a specific exercise at the end of this chapter, and a completed diagram for each mode is included in the answers to this problem set. Working through each mode will generally lead to greater understanding of the modes and how pathways may be integrated.

PROBLEM SET: THE FOUR MODES OF THE PENTOSE PHOSPHATE PATHWAY

IMPORTANT: Please attempt these exercises using your knowledge of the pathways, which will be more meaningful than simply looking at the answers that follow. One purpose of these exercises is to apply one's knowledge of the products of different pathways and how the cell must integrate pathways to produce the products needed. Thus, these exercises further answer the question *when* for these pathways.

INSTRUCTIONS: For each of the 4 modes of the pentose phosphate pathway indicated, draw in the appropriate arrows that will lead to the production of the products needed by the cell in that mode. Note that not all intermediates and reactions are designated on the left for glycolysis/gluconeogenesis.

Exercise 1: Mode 1. Both ribose 5-phosphate and NADPH are needed (Figure 5.17)

(Example situation: Growing tissue—needs ribose 5-phosphate for nucleic acid synthesis and NADPH for reductive biosynthesis.)

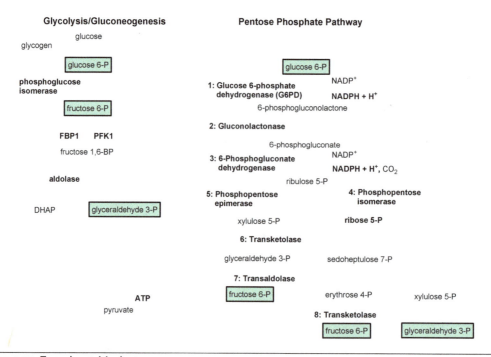

Figure 5.17: Exercise 1. Mode 1.

UNDERSTANDING BIOCHEMICAL PATHWAYS: A PATTERN-RECOGNITION APPROACH

Exercise 2: Mode 2. NADPH only is needed (Figure 5.18)

(Example situation: Adipose tissue—needs NADPH for fatty acid synthesis, but not ribose 5-phosphate.)

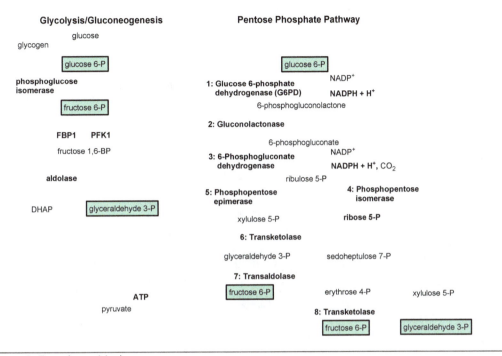

Figure 5.18: Exercise 2. Mode 2.

Exercise 3: Mode 3. NADPH and ATP are needed, but ribose 5-phosphate is not needed (Figure 5.19)

(Example situation: Red blood cell—needs NADPH for maintaining a "reduced atmosphere" and needs ATP for energy, but has no use for ribose 5-phosphate since a red blood cell does not have a nucleus.)

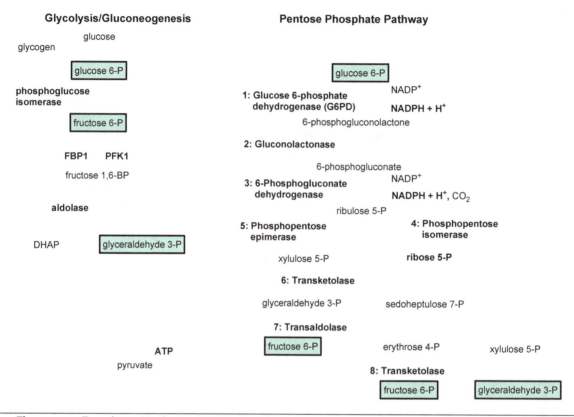

Figure 5.19: Exercise 3. Mode 3.

Exercise 4: Mode 4. Ribose 5-phosphate only is needed (Figure 5.20)

(Example situation: Very rapidly proliferating cell like a tumor cell—needs lots of ribose 5-phosphate but not so much NADPH.)

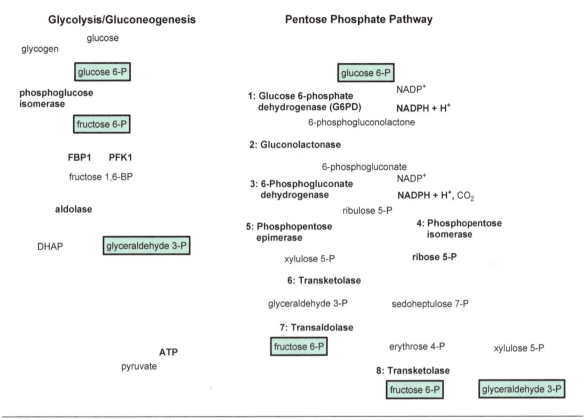

Figure 5.20: Exercise 4. Mode 4.

PROBLEM SET ANSWERS: THE FOUR MODES OF THE PENTOSE PHOSPHATE PATHWAY

Exercise 1: Mode 1. Both ribose 5-phosphate and NADPH are needed (Figure 5.21)
  a. Example: Growing tissue—needs ribose 5-phosphate for nucleic acid synthesis and NADPH for reductive biosynthesis.
  b. Glucose 6-phosphate is formed from glucose entering the cell or glycogen breakdown (depending on the tissue).
  c. Then the first four reactions of the pentose phosphate pathway predominate.
  d. No carbon metabolites are returned to glycolysis.

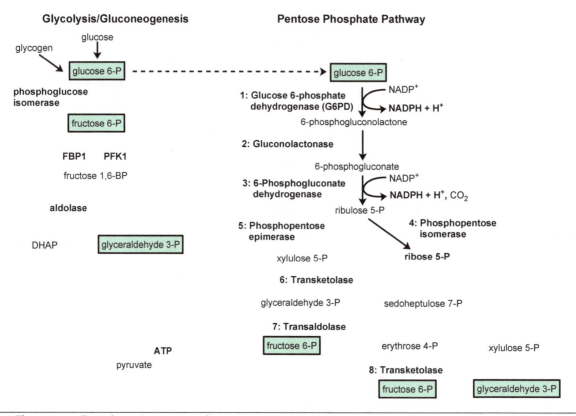

Figure 5.21: Exercise 1. Answer, mode 1.

Exercise 2: Mode 2. NADPH only is needed (Figure 5.22)
   a. Example: Adipose tissue—needs NADPH for fatty acid synthesis, but not ribose 5-phosphate.
   b. An initial pool of glucose 6-phosphate is initially formed from glucose entering the cell or glycogen breakdown (depending on the tissue).
   c. Then do all eight steps of the pentose phosphate pathway.
   d. Recycle the end products of glyceraldehyde 3-phoshpate and fructose 6-phosphate back to re-form *glucose 6-phosphate* via part of the *gluconeogenesis* pathway, then repeat the eight steps of the pentose phosphate pathway.

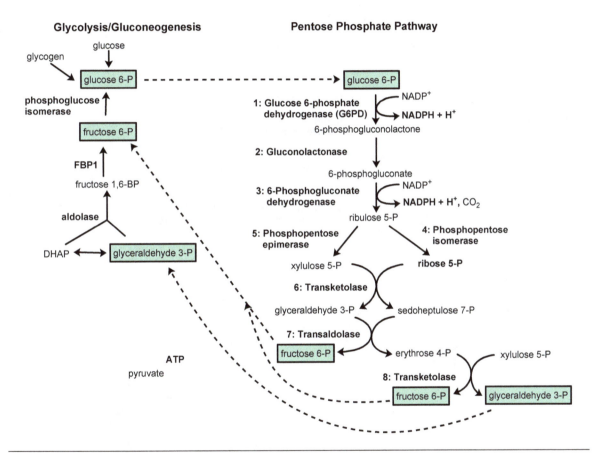

Figure 5.22: Exercise 2. Answer, mode 2.

Exercise 3: Mode 3. NADPH and ATP are needed, but ribose 5-phosphate is not needed (Figure 5.23)
   a. Example: Red blood cell—needs NADPH for maintaining a "reduced atmosphere" and needs ATP for energy, but has no use for ribose 5-phosphate since a red blood cell does not have a nucleus.
   b. Glucose 6-phosphate is formed from glucose entering the cell or glycogen breakdown (depending on the tissue).
   c. Then do all eight steps of the pentose phosphate pathway (though a portion of the glucose 6-phosphate formed may continue directly through glycolysis).
   d. Send the pentose phosphate pathway end products of glyceraldehyde 3-phosphate and fructose 6-phosphate through remainder of *glycolysis* to produce ATP.

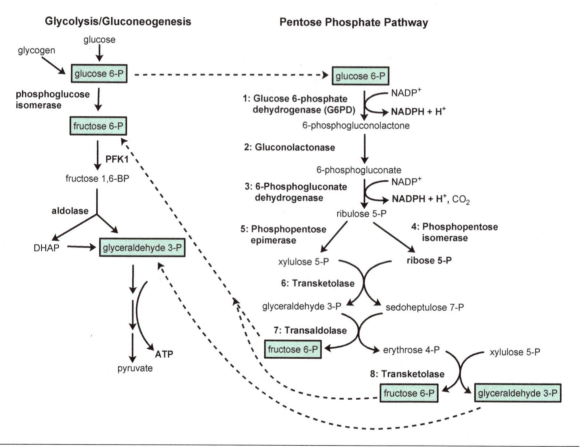

Figure 5.23: Exercise 3. Answer, mode 3.

Exercise 4: Mode 4. Ribose 5-phosphate only is needed (Figure 5.24)

a. Example: Very rapidly proliferating cell like a tumor cell—needs lots of ribose 5-phosphate, but not so much NADPH.
b. Glucose 6-phosphate is formed from glucose entering the cell or from glycogen breakdown (depending on the tissue).
c. Glucose 6-phosphate continues *through glycolysis* to form pools of fructose 6-phosphate and glyceraldehyde 3-phosphate.
d. The fructose 6-phosphate and glyceraldehyde 3-phosphate enter the pentose phosphate pathway *in reverse* (i.e., at reaction 8) to form ribose 5-phosphate.
e. No carbon metabolites will be returned to glycolysis, and no NADPH will be formed.

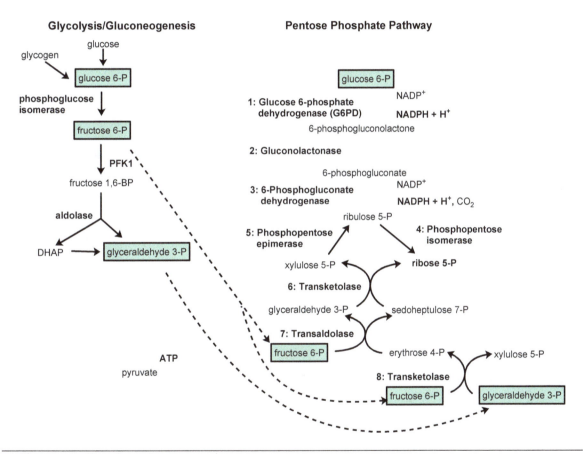

Figure 5.24: Exercise 4. Answer, mode 4.

CHAPTER 6

β-OXIDATION OF FATTY ACIDS AND KETONE BODY SYNTHESIS

OBJECTIVES

1. Define the process of β-oxidation of fatty acids (a.k.a. β-oxidation or fatty acid oxidation).
   a. Identify the starting and ending products of the pathway, including their structures.
2. Explain the purpose of the pathway of β-oxidation.
   a. Identify all the products of the pathway, the enzymes that produce them, and for what the products can be used.
3. Identify where β-oxidation takes place in the cell and what tissues can carry out this pathway.
   a. Identify the sources of free fatty acids and how they are transported in the blood.
   b. Explain how acyl CoA is shuttled into the mitochondria and the role of carnitine in the process.
4. Explain how β-oxidation is carried out in the cell.
   a. Describe the four repeated steps of β-oxidation and identify the enzymes and cofactors involved in these steps.
   b. Calculate the number of cycles needed to oxidize fatty acids of even- and odd-numbered chain lengths and the total number of products produced (i.e., the number of acetyl CoA, propionyl CoA, $FADH_2$, and NADH produced for the complete oxidation of the fatty acid in question).
5. Explain when β-oxidation takes place.
   a. Identify the regulatory enzymes of β-oxidation; identify what is the primary regulator of CAT I and how hormones regulate β-oxidation.
6. Define the process of ketone body synthesis.
   a. Identify the starting and ending products of the pathway, including their structures.
7. Explain the purpose of the pathway of ketone body synthesis.
8. Identify where ketone body synthesis takes place in the cell and what tissue(s) can carry out this pathway.
9. Explain how ketone body synthesis is carried out in the cell and how ketone bodies are utilized by other tissues.
   a. Describe the steps of ketone body synthesis, the enzymes involved, and explain the fates of the three ketone bodies produced.
10. Explain when ketone body synthesis takes place (i.e., under what conditions ketone bodies are produced).
    a. Explain the regulation by acetyl CoA concentrations that coordinate fatty acid oxidation, ketone body synthesis, and gluconeogenesis.
    b. Explain why starvation increases ketone body production.

OVERVIEW OF LIPID METABOLISM

Lipids are an important source of fuel for the body. One pound of fat (triacylglycerols) will yield about 4,000 kcal. One pound of glycogen, the storage form of glucose, will yield about 400 kcal. As one understands the basic recipe of metabolism, this makes sense. The more oxidation steps that must be done to oxidize carbons to carboxylic acids, the more $FADH_2$ and NADH are produced. The $FADH_2$ and NADH produced are used by the electron transport chain for the production of ATP. A sixteen-carbon fatty acid (i.e., palmitate) has fifteen carbons that are at the level of an alkane and one carbon already at the level of a carboxylic acid. To oxidize all fifteen of these carbons up to carboxylic acids, so they can be clipped off as CO_2 to burn them off, will produce a lot of $FADH_2$ and NADH for ultimate ATP production. Glucose, on the other hand, is only six carbons. Of these six carbons, five are at the level of an alcohol and one is at the level of an aldehyde. Glucose has fewer carbons, and they are already more oxidized than a fatty acid. Therefore, the cell has to do less oxidation reactions to oxidize these carbons to carboxylic acids, and less $FADH_2$ and NADH are produced. Fatty acids, then, are the body's major storage form of energy, and they are stored as triacylglycerols (commonly abbreviated as TAGs, or TGs for triglycerides).

Another benefit to storing triacylglycerols over glycogen is that triacylglycerols are hydrophobic, meaning they do not like water. A triacylglycerol molecule consists of three fatty acids linked through ester bonds to glycerol, as shown in **figure 6.1**. Glycogen, though, is hydrophilic and has water associated with it. By storing more fuel as triacylglycerols than as glycogen, the body will not have excess weight due to water associated with glycogen.

The tissues of the body must always have ready access to a fuel source that is circulating through the blood, which the author terms "fast food." Lipids, though, are not our fast food, or readily available circulating fuel source. The hydrophobicity of lipids makes them a great storage form of fuel but not a circulating fuel source. Because lipids are hydrophobic, they are not easily transported in the blood, which largely consists of water. Maintaining a circulating supply of lipids in the blood is more difficult, as lipids like triacylglycerols and cholesterols must be packaged in special vesicles called lipoprotein particles (i.e., chylomicrons, VLDLs, LDLs, and HDLs). Individual fatty acid molecules must be attached to a carrier protein, albumin, for transport through the blood.

Thus, triacylglycerols account for most of our stored energy, but it is not the body's primary "circulating fuel" because it hates water. Glucose does like water, as it is hydrophilic and water soluble. A cell cannot get as much energy from a molecule of glucose, but it is easier to circulate in the blood. Therefore, glucose serves as the primary water-soluble fuel source that is readily available (i.e., fast food) for use by tissues. Ketone bodies, which are water-soluble derivatives of fatty acids, are a second major circulating fuel that will be also be discussed in this chapter.

Lipids are also important structural components of biological membranes. These lipids include phospholipids (i.e., glycerophospholipids) shown in **figure 6.1**, sphingolipids, glycolipids, and cholesterol. Membranes

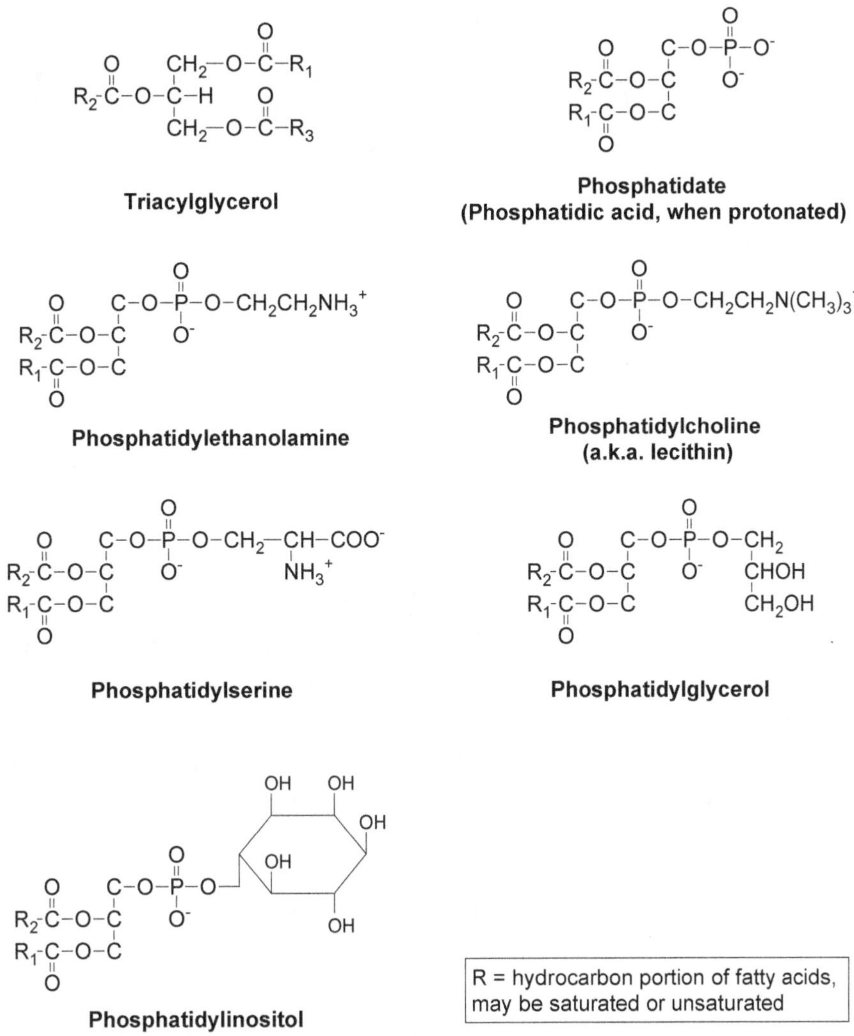

Figure 6.1: Basic structure of a triacylglycerol and the structures of the common glycerophospholipids.

are lipid bilayers. Membranes form hydrophobic barriers and are responsible for compartmentation of cells. Lipids that reside in membranes must be amphipathic, meaning they have a hydrophilic (polar) portion and a hydrophobic (nonpolar) portion of the molecule. The hydrophilic region can interact with water, and the hydrophobic portion will form the hydrophobic center of a membrane bilayer. Reciprocal pathways often take place in separate compartments of a cell, or at least some portion of one of the pathways occurs in a different compartment. As mentioned previously, many catabolic pathways are in the mitochondria, while synthetic pathways are often in the cytosol (though there are exceptions in both compartments). Some lipid components of membranes play important roles in signal transduction pathways. Other lipids serve as hormones (i.e., steroid hormones, derived from cholesterol; and the eicosanoid hormones, which are derived from arachidonic acid shown in **table 6.1**).

Table 6.1: Fatty acids of importance to humans

NUMERICAL SYMBOL	STRUCTURE	TRIVIAL NAME	SYSTEMATIC NAME
16:0	$CH_3-(CH_2)_{14}-COOH$	Palmitic acid	Hexadecanoic
18:2(9,12)	$CH_3-(CH_2)_3-(CH_2-CH=CH)_2-(CH_2)_7-COOH$	Linoleic acid	*cis,cis*9,12-Octadecatrienoic
18:3(9,12,15)	$CH_3-(CH_2-CH=CH)_3-(CH_2)_7-COOH$	Linolenic acid	*cis,cis,cis*-9,12,15-Octadecatrienoic
20:4(5,8,11,14)	$CH_3-(CH_2)_3-(CH_2-CH=CH)_4-(CH_2)_3-COOH$	Arachidonic acid	*cis,cis,cis,cis*-5,8,11,14-Icosatetraenoic

Certain lipids must be obtained in the diet. There are four lipid-soluble vitamins (vitamins A, D, E, and K), which are important to obtain in the diet. These vitamins carry out important biological functions, of which a few functions are described here. Vitamin A derivatives are visual pigments, as well as hormones. Vitamin A is commonly found in fish liver oils, liver, eggs, whole milk, and butter. The yellow vegetables, such as carrots and sweet potatoes, provide β-carotene that can also be converted to vitamin A. Vitamin D derivatives are hormones for the regulation of calcium levels, which is important for bone formation. Vitamin D_3 is formed in skin from a cholesterol derivative using ultraviolet rays of light. Most people do not get sufficient amounts naturally, so similar vitamin D derivatives are added to milk and juice products. Vitamin E derivatives serve as antioxidants, and vitamin K derivatives are important in the blood-clotting cascade. Vitamin E is abundant in wheat germ and can also be obtained from eggs and vegetable oils. Forms of vitamin K are obtained from green leafy vegetables and some intestinal bacteria.

Linoleic and linolenic acids are two essential fatty acids for humans (see **table 6.1**). These two fatty acids must also be obtained in the diet. These are the omega-6 (ω-6) and omega-3 (ω-3) fatty acids that one commonly hears about regarding dietary supplements. Recall that the "omega" carbon is the methyl end carbon of a fatty acid (see **figure 1.1**). In counting in from the methyl end the molecule, an omega-6 fatty acid will have its first double bond six carbons in, while an omega-3 fatty acid has its first double bond three carbons in from the methyl end of the molecule.

While the synthesis of complex lipid structures (i.e., triacylglycerols, phospholipids, lipid hormones, etc.) will not be covered in this textbook, understanding what fatty acids are used for in the cell is important. Membrane lipids (i.e., phospholipids, sphingolipids, and glycolipids), triacylglycerols, and lipoproteins all have fatty acids as components. Thus, knowing the catabolism of what molecules serve as the source of fatty acids for β-oxidation, as well as the possible fates of a newly synthesized fatty acid, is an important level of understanding for the lipid pathways covered in this text. A cell will not, though, synthesize a new fatty acid only to turn around and break it down in β-oxidation, as that would be a futile cycle. Fatty acid synthesis and β-oxidation, as reciprocal pathways, are regulated such that both of these pathways are not on at the same time in the same cell.

β-OXIDATION OF FATTY ACIDS (A.K.A. FATTY ACID OXIDATION)

The name of this pathway is its definition (*what*). β-oxidation of fatty acids, also known simply as *β-oxidation* or as *fatty acid oxidation*, is the breakdown or catabolism of fatty acids via oxidation at the β-carbon to form molecules of acetyl CoA. In this pathway, the starting material is a fatty acid in which carbon 3 (the β-carbon) is at the level of an alkane. The β-carbon of the fatty acid will need to be oxidized to a ketone, prior to clipping off a molecule of acetyl CoA.

The purpose (*why*) of β-oxidation of fatty acids is to produce reducing power in the form of $FADH_2$ and NADH that will donate their electrons directly to the electron transport chain for ATP production. The acetyl CoA can be sent into the TCA cycle to further catabolize the two-carbon acetyl unit to CO_2 and produce more reducing power to be used for ATP production. In the liver, the acetyl CoA can also be used for ketone body production.

This pathway takes place (*where*) in the mitochondrial matrix. However, the free fatty acids are in the cytosol, and the cell needs to move them into the matrix. Therefore, a transport step is needed for this pathway. Most tissues can take up fatty acids circulating in the blood for fuel, except for mature red blood cells and the brain. Red blood cells cannot use fatty acids, because they lack mitochondria. The brain cannot import fatty acids from the blood because fatty acids cannot cross the blood-brain barrier. Neurons, though, can catabolize fatty acids from their own cellular membrane turnover.

The overview of the pathway of β-oxidation of fatty acids is shown in **figure 6.2**. For learning the process (*how*) of β-oxidation, one needs to know the sources of fatty acids and how they are transported into the mitochondrial matrix. The four repeated steps of β-oxidation follow the principles of the oxidation states flowchart (**figure 1.7**) for the series of reactions required to put a keto group on a carbon. In fact, exercise 4 at the end chapter 1 required one to put in order the correct sequence of molecules for the first three reactions of β-oxidation based on their structures, as well as name the enzymes for these reactions.

Sources of the fatty acids include dietary fats, stored body fat (i.e., triacylglycerols in adipose cells), and membrane lipid turnover. Membranes are constantly being damaged and repaired. Membranes are also newly synthesized as cells grow and divide. Thus, the degradation of membrane components such as phospholipids, sphingolipids, or glycolipids can provide fatty acids for β-oxidation.

Lipases are enzymes that cleave components from the complex lipids such as phospholipids or triacylglycerols. There are specific types of lipases that act on specific classes of complex lipids. *Hormone-sensitive lipase* is a hormonally controlled lipase that clips fatty acids off stored triacylglycerol molecules. Several types of *phospholipases* cleave the fatty acids and the headgroups from the various phospholipids. The pancreas secretes lipases for the digestion of various dietary lipids.

The transport of lipids through the blood is a complex process due to the hydrophobicity of lipids. As previously mentioned, the larger complex lipids are transported in the blood in vesicles called lipoprotein

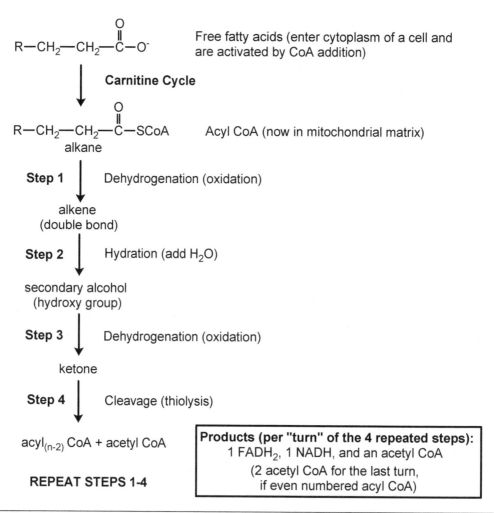

Figure 6.2: Overview of β-oxidation of fatty acids.

particles, which have a monolayer of phospholipids and specific proteins. The hydrophobic interior of these lipoprotein particles allow triacylglycerols and cholesterol esters to be transported inside the vesicle. Individual fatty acid molecules must also be transported in the blood using a carrier. The protein albumin is the carrier of individual fatty acids through the blood.

Individual or "free" fatty acids are not components of membranes, and they must always be attached to a carrier molecule inside or outside of a cell. Free fatty acids act as detergents and will break up membranes causing a cell to lyse. As fatty acids are brought into a cell or synthesized de novo, they are quickly attached to a carrier molecule, usually coenzyme A.

The attachment of a fatty acid to a cellular carrier requires an input of energy, referred to as activation of a free fatty acid. This input of energy is similar to the input of energy used in glycolysis in the beginning of the pathway. The activation step is the attachment of a fatty acid to coenzyme A (CoA) using ATP, as shown

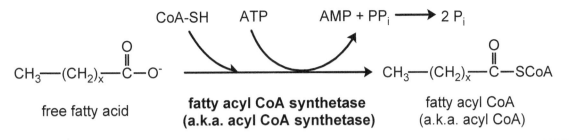

Figure 6.3: Activation of a free fatty acid.

in **figure 6.3**. The product of this reaction is a fatty acyl CoA, or more simply referred to as an acyl CoA. This activation step occurs by a family of isozymes (each isozyme activates certain fatty acid chain lengths) on the outer mitochondrial membrane.

The name of the family of isozymes for this step indicates the reaction they carry out. An enzyme that synthesizes bonds between two molecules to form larger products is called either a *synthase* or a *synthetase*. A synthase does not require energy, while a synthetase does require energy. Both synthases and synthetases are named for the *product* of the reaction. In this reaction ATP is used to attach a fatty acid to the free sulfhydryl group of coenzyme A, forming a thioester bond between the two molecules. The product of the reaction is an acyl CoA. Therefore, the name of the family of enzymes that catalyzes this reaction is *fatty acyl CoA synthetase*, or simply *acyl CoA synthetase*. The various isozymes of the acyl CoA synthetases work on fatty acids that are of short, intermediate, or long carbon chains.

In the reaction (shown in **figure 6.3**), two phosphates of ATP are cleaved off to form AMP plus a molecule called pyrophosphate (PP_i). Pyrophosphate is immediately cleaved by an enzyme called *pyrophosphatase* to form two inorganic phosphate groups. The cleavage of pyrophosphate drives the reaction forward and makes this reaction *irreversible*. This cleavage of two high-energy phosphate bonds is the equivalent of two ATP forming two ADP molecules, which will become important when calculating ATP yields from the complete catabolism of a fatty acid.

At this point the fatty acid is in the cytosol of the cell safely attached to coenzyme A. This acyl CoA does not have to be transported into the mitochondrial matrix for further catabolism. It could be used for membrane lipids, if the cell needs to repair membranes or synthesize new membranes; for storage on a triacylglycerol molecule; for the formation of a lipoprotein; and so forth. If the cell does need to catabolize the fatty acid further via β-oxidation, the acyl CoA does need to be transported into the mitochondrial matrix, as will be described next.

The fatty acid was activated to an acyl CoA on the outer mitochondrial membrane, and the cell now needs to get it into the mitochondrial matrix. The carnitine cycle is the process by which the cell transports the acyl CoA molecules into the mitochondrial matrix for β-oxidation. Recall that the inner mitochondrial membrane is very impermeable. Special transporters are required to move molecules across the inner mitochondrial membrane. The process of moving the fatty acid into the matrix requires two enzymes and a

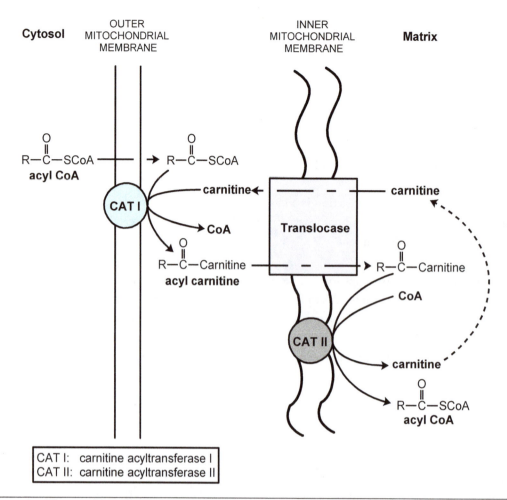

Figure 6.4: The carnitine shuttle.

transporter. The two enzymes are carnitine acyltransferase I (CAT I) and carnitine acyltransferase II (CAT II); see **figure 6.4**. CAT I is a regulated enzyme located in the outer mitochondrial membrane. CAT I exchanges the coenzyme A on an acyl CoA for carnitine to form acyl carnitine. The formation of the acyl carnitine by CAT I commits the fatty acid for movement into the mitochondria for further catabolism.

The transporter is called the acyl carnitine/carnitine translocase (or transporter). When a transporter in the inner mitochondrial membrane opens, it typically lets a molecule into the matrix and lets something out. In this case the translocase transports acyl carnitine into the matrix, and free carnitine is transported out. Once the acyl carnitine is in the matrix, CAT II (located in the inner mitochondrial membrane) clips the carnitine molecule off and attaches the fatty acid back onto a coenzyme A molecule—re-forming an acyl CoA that is now located in the mitochondrial matrix for catabolism by the β-oxidation pathway. CAT II is not regulated.

Carnitine can be obtained from dietary sources, primarily in meat. Carnitine can also be synthesized from lysine and methionine, but only in the liver and kidneys. Other tissues, especially the heart and skeletal muscles, depend on a supply of carnitine from endogenous synthesis in the liver and kidneys or from the diet. Carnitine from either of these sources must be circulated in the blood for uptake by these tissues. Skeletal muscle, of particular note, contains over 95 percent of all of the carnitine in the body. Strict vegetarians and vegans may need to ensure they are getting plenty of carnitine from alternative dietary sources.

Exercise 4 at the end chapter 1 puts in order the first three steps of the four repeated steps of β-oxidation. For β-oxidation, a keto group needs to be put on a carbon that is initially at the level of an alkane. This pathway proceeds from the top of the oxidation states flowchart down to the keto group (shown in **figure 1.7**). This is the exact same sequence of three reactions seen in the TCA cycle to go from succinate to oxaloacetate. In exercise 4 of chapter 1, the molecules were named using the first letter of their molecular name. The goal of β-oxidation of fatty acids is to produce acetyl CoA and reducing power in the form of $FADH_2$ and NADH. The starting material shown in **figure 6.5** is an acyl CoA—specifically palmitoyl CoA, a sixteen-carbon saturated fatty acid attached to coenzyme A. Palmitate is one of the more common fatty acids in humans. If the cell simply clipped off the first two carbons of the fatty acid attached to the coenzyme A, a molecule of acetyl CoA would be produced. However, the other product would be a hydrocarbon, a molecule with only carbons and hydrogens. Hydrocarbons are very hydrophobic and unreactive. A cell would not be able to do anything with a hydrocarbon product. Therefore, the four-step reaction sequence of the pathway of β-oxidation of fatty acids is set up to yield a molecule of acetyl CoA and an acyl CoA molecule that is two carbons shorter than the one used in the first reaction. The shorter fatty acid can then continue to be catabolized in subsequent rounds of these four steps. The sequence of the four repeated steps of β-oxidation of fatty acids is shown in **figure 6.5**.

The name "β-oxidation of fatty acids" indicates exactly what happens in this pathway. The β-carbon (i.e., carbon 3) of the fatty acid is oxidized. Specifically, the β-carbon is oxidized to a keto group. In **figure 6.5** carbon 3, the β-carbon, is at the level of an alkane in palmitoyl CoA (or any acyl CoA). Following the oxidation states flowchart (**figure 1.7**), an alkane can be oxidized to a keto group by first oxidizing to an alkene, then hydration to a secondary alcohol, followed by oxidation to a keto group. The first three steps of β-oxidation are these exact three steps.

In step 1 the acyl CoA (palmitoyl CoA for this example) is oxidized to an enoyl CoA. Enoyl CoA is the generic name for any fatty acid containing a double bond (i.e., an unsaturated fatty acid). To be more specific, the double bond introduced in this step is a *trans* double bond. Double bonds introduced into fatty acids that are incorporated into membrane lipids are typically in the *cis* configuration for structural reasons. In this reaction, though, the *trans* configuration of the double bond is fine because the double bond will be removed in the next step. Recall from chapter 1 when oxidizing from an alkane to an alkene, FAD is the coenzyme used for the coupled reduction reaction, as shown in **figure 6.5**.

Enzymes that carry out oxidation-reduction reactions using NAD^+ or FAD are called *dehydrogenases*. Dehydrogenases are named for the *more reduced* molecule. Therefore, the name of the enzyme for reaction 1 is *fatty acyl CoA dehydrogenase*, or more simply, *acyl CoA dehydrogenase*. In the mitochondria there is a family of different acyl CoA dehydrogenases (i.e., isozymes) that that have distinct, yet overlapping, specificities

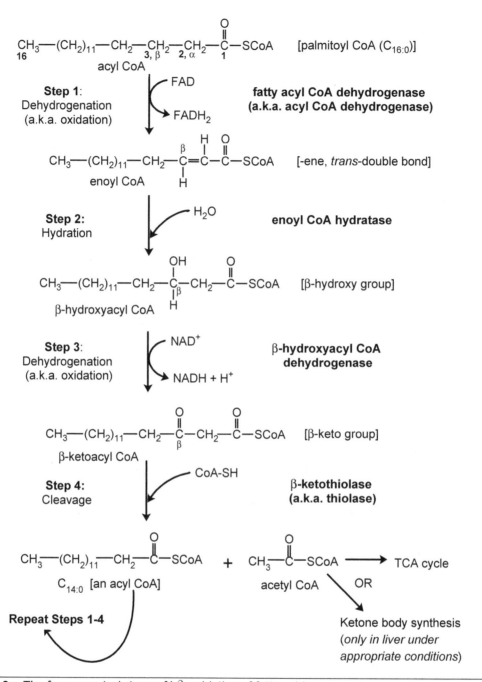

Figure 6.5: The four repeated steps of bβ-oxidation of fatty acids.

for fatty acids of particular chain lengths. For example, long-chain acyl CoA dehydrogenase (LCAD) works on fatty acids of about twelve to eighteen carbons, medium-chain acyl CoA dehydrogenase (MCAD) works on fatty acids of about four to fourteen carbons, while short-chain acyl CoA dehydrogenase (SCAD) acts on fatty acids of about four to eight carbons. As a clinical note, there are known genetic disorders due to mutations in the genes for each of these isozymes.

In step 2 (**figure 6.5**), water is added across the double bond of the enoyl CoA such that the hydroxy group is placed on the β-carbon (carbon 3), forming a secondary alcohol group. The product of this reaction is called β-hydroxyacyl CoA (or 3-hydroxyacyl CoA). This reaction adds water to the molecule, so the enzyme is called a *hydratase*. The enzyme is named after the molecule that is accepting the water molecule, enoyl CoA. Thus, the enzyme is called *enoyl CoA hydratase*.

In step 3 (**figure 6.5**), the secondary alcohol of β-hydroxyacyl CoA is oxidized to a ketone forming β-ketoacyl CoA. This is an oxidation reaction, so a reduction reaction must simultaneously occur. NAD^+ is reduced to $NADH + H^+$, as is typical for oxidations of alcohols to ketones. Enzymes that carry out oxidation-reduction reactions using NAD^+ or FAD are called *dehydrogenases*. Dehydrogenases are named for the *more reduced* molecule, which is β-hydroxyacyl CoA. The enzyme that catalyzes reaction 3 is thus named *β-hydroxyacyl CoA dehydrogenase*.

At this point a keto group is now on the β-carbon (carbon 3) of a fatty acid. Hopefully, the three reactions needed to accomplish this make sense in the context of the oxidation states flowchart (or basic recipe of metabolism). In step 4 a molecule of acetyl CoA is cleaved off using another molecule of coenzyme A to also produce a two-carbon shorter acyl CoA (a fourteen-carbon fatty acyl CoA, as shown in **figure 6.5**). The enzyme that catalyzes this reaction is called *β-ketothiolase* (or just *thiolase*). By putting the keto group on the β-position of the fatty acid, a molecule of coenzyme A with is free sulfhydryl (or thiol) group is now used to cleave the bond (i.e., thiolase) between carbons 2 and 3. The resulting attachment of the new molecule of CoA to carbon 3 (the keto group carbon) produces a new shorter acyl CoA, along with a molecule of acetyl CoA. The fourteen-carbon acyl CoA can continue to be catabolized through subsequent rounds of these four steps of β-oxidation.

Calculation of the net ATP from β-oxidation of a fatty acid

For calculating how much ATP can be produced from the catabolism of a fatty acid, one first needs to figure out the products of the complete catabolism of a given fatty acid. The first step is to determine how many acetyl CoA molecules can be formed from the catabolism of a particular fatty acid. The next step is to determine how many cycles of these four repeated steps are needed to produce those molecules of acetyl CoA. If the starting fatty acid has an even number of carbons, just divide the number by two (because the acetyl unit is two carbons). For the sixteen-carbon fatty acid shown in **figure 6.5**, a total of eight acetyl CoA molecules can be produced. Seven cycles of the four repeated steps are needed to produce those eight acetyl CoA molecules. Only seven cycles of the steps are needed because the last round is starting with a four-carbon fatty acid, which forms the last two acetyl CoA molecules when it is cut in half by the thiolase. The seven cycles of the four repeated steps will also yield seven $FADH_2$ molecules and seven NADH molecules. Since this pathway occurs in the mitochondrial matrix, the $FADH_2$ and NADH molecules can deliver their electrons directly to the electron transport chain. The eight acetyl CoA molecules will now "turn" the TCA cycle eight times, producing even more reducing power and some GTP, while completely catabolizing the fatty acids to CO_2.

Table 6.2: Net energy (ATP) yield from the complete catabolism of palmitoyl CoA (C_{16})

PRODUCTS OF β-OXIDATION	FROM TCA CYCLE	ATP YIELD
8 acetyl CoA	(8 x) 3 NADH (x 2.5)	60
	(8 x) 1 $FADH_2$ (x 1.5)	12
	(8 x) 1 GTP (x 1)	8
7 $FADH_2$ (x 1.5)		10.5
7 NADH (x 2.5)		17.5
		108
2 high-energy PO_4^{2-} bonds broken during activation stage (ATP to AMP + PP_i; PP_i to $2P_i$)		−2
		106 ATP (net yield)

Table 6.2 shows the net energy (ATP) yield from the complete oxidation of palmitoyl CoA to sixteen molecules of CO_2. Recall that an NADH yields about 2.5 ATP, and an $FADH_2$ yields about 1.5 ATP from the electron transport chain. As one can see from the table, about 106 ATP can be produced from the catabolism of a fatty acid, which is much more than a molecule of glucose (30 to 32 ATP). The catabolism of a fatty acid produces much more reducing power (NADH and $FADH_2$) formed during the oxidation of the carbons from the level of an alkane up to a ketone in the four repeated steps, and then to CO_2 in the TCA cycle.

Catabolism of other types of fatty acids

The body contains more than just sixteen-carbon saturated fatty acids, both obtained in the diet and from de novo synthesis. There are odd-numbered carbon chain length fatty acids, branched chain fatty acids, and unsaturated fatty acids. Unsaturated fatty acids are fatty acids that have one or more double bonds, usually in the *cis* configuration. *Cis* double bonds introduce a structural change, termed "kinks," in the hydrocarbon portion of the fatty acid. These *cis* double bonds prevent these types of fatty acids, particularly on membrane lipids, from packing together as tightly and allowing for membrane fluidity.

The enoyl CoA hydratase (for reaction 2) can only hydrate an enoyl CoA that has a *trans* double bond between carbons 2 and 3. During rounds of the four repeated steps of β-oxidation of an unsaturated fatty acid a *cis* double bond is typically encountered either between carbons 2 and 3 or between carbons 3 and 4. In either of these cases, the enzyme *enoyl CoA isomerase* isomerizes the double bond to a *trans* double bond and, if necessary, moves it between carbons 2 and 3. An $FADH_2$ molecule will not be formed for that round of repeated steps because the molecule already has a double bond (i.e., enoyl CoA isomerase catalyzes the first step of that round, rather than the acyl CoA dehydrogenase enzyme).

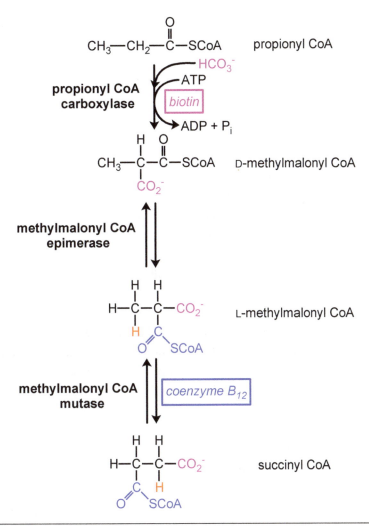

Figure 6.6: Conversion of propionyl CoA, from β-oxidation of odd-numbered carbon fatty acids, to succinyl CoA for complete oxidation by the TCA cycle.

For fatty acids that have an odd-numbered carbon chain length (i.e., a seventeen-carbon fatty acid), the last round of four repeated steps starts with a five-carbon-long fatty acid. In the thiolase step, an acetyl CoA molecule is produced along with a three-carbon acyl CoA called propionyl CoA. The cell still needs to further catabolize this molecule by getting it into the TCA cycle. The propionyl CoA is converted to succinyl CoA in three reactions, as shown in **figure 6.6**. The first enzymatic reaction requires ATP and CO_2 (from bicarbonate, HCO_3^-), to carboxylate the propionyl CoA to a four-carbon molecule called D-methylmalonyl CoA. The enzyme that catalyzes this reaction is *propionyl CoA carboxylase*. It requires biotin and carries out a similar mechanism as other carboxylases (i.e., pyruvate carboxylase, discussed in chapter 5; and acetyl CoA carboxylase, discussed in chapter 7). The second reaction is catalyzed by *methylmalonyl CoA epimerase*, which isomerizes D-methylmalonyl CoA to L-methylmalonyl CoA. The third reaction is catalyzed by *methylmalonyl CoA mutase* and requires a coenzyme form of vitamin B_{12}. This reaction is unusual in that a hydrogen and the carbonyl CoA group on adjacent carbons exchange positions to form succinyl CoA, which can then enter the TCA cycle. The propionyl CoA molecule will not yield as much reducing power as an acetyl CoA molecule because it enters the TCA cycle further into the cycle. Note that the formation of succinyl CoA from propionyl CoA is the only gluconeogenic precursor derived from the catabolism of fatty acids because succinyl CoA enters the TCA cycle after the reactions where the two molecules of CO_2 are produced.

There are also other types of fatty acid catabolic processes located in other organelles, which are briefly described here. In peroxisomes, β-oxidation works primarily on very long carbon chain length fatty acids (i.e., hexacosanoic acid, a $C_{26:0}$) and branched chain fatty acids such as phytanic acid and pristanic acid. Branched chain fatty acids typically have a methyl group in the β-carbon position, which does not allow for oxidation of the β-carbon. Therefore, catabolism of the branched chain fatty acids in the peroxisomes uses a process called α-oxidation. In α-oxidation, the branched fatty acid is oxidized to a hydroxyl group on the α-carbon. This product is then shortened by one carbon by decarboxylation, forming an aldehyde. The aldehyde is subsequently oxidized to a carboxylic acid. Peroxisomes catabolize fatty acids to shorter carbon chain lengths, which are then transported to the mitochondrial matrix for complete oxidation.

Another process for fatty acid catabolism is ω-oxidation, which is oxidation of the ω-carbon (the methyl end). This type of fatty acid catabolism occurs in the endoplasmic reticulum and is normally a relatively minor pathway of fatty acid catabolism in humans (and other mammals). However, fatty acid catabolism by ω-oxidation is up-regulated in individuals with a defect in an enzyme of β-oxidation or in those who have a carnitine deficiency.

Regulation of β-oxidation of fatty acids

There are several enzymes involved in the regulation of β-oxidation of fatty acids, including the lipases that generate the sources of fatty acids, as well as CAT I involved in transport of fatty acids to the matrix. Hormone-sensitive lipase, which is an important enzyme in clipping off fatty acids from triacylglycerols in adipose tissue, is hormonally regulated. Insulin (high blood sugar indicator) and glucagon (low blood sugar) are two important hormones that regulate this lipase. Glucagon stimulates hormone-sensitive lipase in adipose tissue. The activation of this lipase provides fatty acids that will be transported via albumin to the liver (one of the main target tissues of glucagon) to be used for ketone body synthesis. In the fed state (insulin levels rise), excess nutrients are stored as triacylglycerols. When insulin levels rise, the hormone-sensitive lipases are inhibited because fatty acid synthesis and triacylglycerol synthesis are occurring in the cells to store these nutrients.

The key step of regulation of β-oxidation of fatty acids, though, occurs at the point of transport into the mitochondrial matrix. Fatty acids, as well as being an important fuel source, are also structures of membrane lipids, lipoproteins, and so on. Once a fatty acid is transported into the mitochondrial matrix, the fatty acid is essentially committed to the oxidative fate (i.e., catabolism). As mentioned previously, CAT I is a regulated enzyme. CAT I is inhibited by high concentrations of malonyl CoA, which is a substrate for fatty acid synthesis. The transport of fatty acids into the mitochondrial matrix (of which the first step is carried out by CAT I) will be inhibited if fatty acid synthesis is occurring in the cell. The cell does not expend energy synthesizing a fatty acid, only to turn around and break it down by β-oxidation. That would be a futile cycle. Therefore, these reciprocal pathways are not on at the same time in the same cell, and the regulation of CAT I is one regulation step that ensures this does not happen.

Two of the enzymatic steps of β-oxidation are also regulated: β-hydroxyacyl CoA dehydrogenase and β-ketothiolase. The ratio of NADH/NAD$^+$ regulates β-hydroxyacyl CoA dehydrogenase. When this ratio is high, NADH predominates indicating the cell has plenty of energy (i.e., the electron transport chain is inhibited due to plenty of ATP), which inhibits this enzyme. High concentrations of acetyl CoA inhibits β-ketothiolase, which is another indicator of sufficient energy in the cell.

KETONE BODY SYNTHESIS

Ketone body synthesis (*what*) is the synthesis of water-soluble derivatives of lipids. The cell catabolizes fatty acids (via β-oxidation) to form acetyl CoA, which will be used to form water-soluble derivatives called ketone bodies. Of the three molecules referred to as ketone bodies, only two of them actually have a keto group, as will be discussed further.

The purpose of the production of ketone bodies (*why*) is that they are used in proportion to their blood concentration by extra hepatic tissues. Skeletal and cardiac muscle use them, and the renal cortex has a preference for them. During starvation or diabetic conditions, ketone bodies become a major fuel source for the brain. Since fatty acids cannot cross the blood-brain barrier, ketone bodies are water-soluble derivatives of fatty acids that can cross the blood-brain barrier. The brain prefers glucose, though, over ketone bodies. This is one reason why maintaining glucose in the blood is a metabolic priority. The brain will use ketone bodies, but it usually requires about forty-eight hours of glucose starvation before the brain will begin to use them. Note that mature red blood cells cannot use ketone bodies. Cells that utilize ketone bodies convert them back to two molecules of acetyl CoA in the mitochondria. Since mature red blood cells do not have mitochondria, they cannot use ketone bodies as a fuel source.

Ketone body synthesis takes place (*where*) in the liver mitochondrial matrix. Only the liver can do ketone body synthesis. While the liver synthesizes the ketone bodies, it cannot use them, because it lacks the enzyme needed to convert acetoacetate to acetoacetyl CoA. The lack of this enzyme prevents the liver from doing a futile cycle of synthesis and breakdown of ketone bodies.

In ketone body synthesis (*how*), all of the carbons of the ketone bodies come from molecules of acetyl CoA. While the ketone bodies are three or four carbons, their synthesis proceeds through a six-carbon activated intermediate, 3-hydroxy-3-methylglutaryl CoA (HMG-CoA). There are three ketone bodies, and **figure 6.7** shows the pathway of ketone body synthesis.

The pathway of ketone body synthesis is three to four steps, depending on the ketone body produced. Reaction 1 (**figure 6.7**) starts with the condensation of two molecules of acetyl CoA. This reaction is

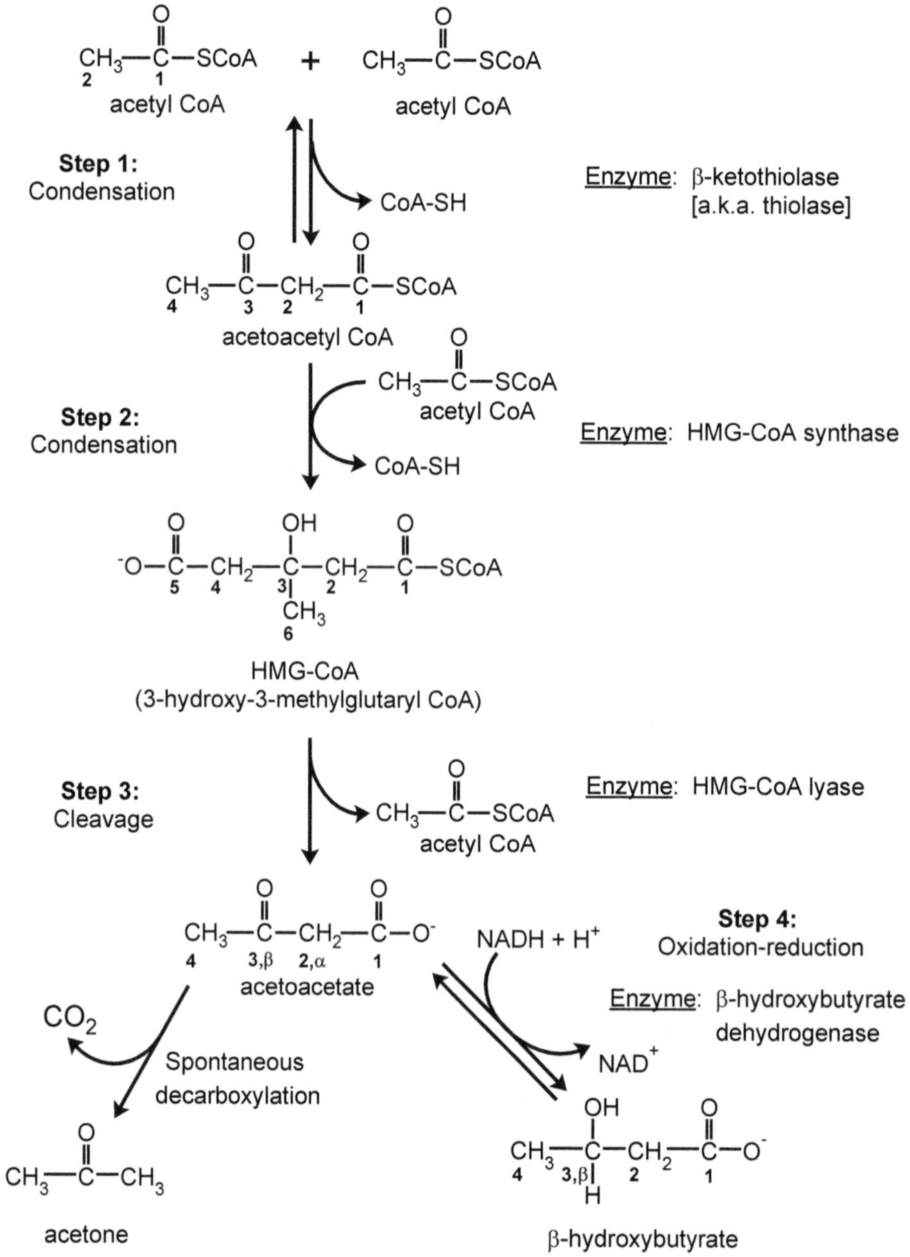

Figure 6.7: Pathway of ketone body synthesis (three to four steps).

catalyzed by *β-ketothiolase*, the same enzyme that carried out step 4 of β-oxidation. For ketone body synthesis, β-ketothiolase carries out the reverse reaction condensing two molecules of acetyl CoA (carbon 1 of one molecule is attached to carbon 2 of the second molecule) to produce acetoacetyl CoA. The name "acetoacetyl CoA" indicates its structure: two acetate molecules connected together end to end and attached to coenzyme A.

146   UNDERSTANDING BIOCHEMICAL PATHWAYS: A PATTERN-RECOGNITION APPROACH

The pathway goes through a six-carbon intermediate using a third acetyl CoA molecule. In reaction 2 (**figure 6.7**), this third molecule of acetyl CoA is condensed at its methyl carbon to carbon 3 of acetoacetyl CoA. The removal of the coenzyme A from this third molecule of acetyl CoA drives the reaction forward to form the six-carbon product 3-hydroxy-3-methylglutaryl CoA (HMG-CoA). The full name of the molecule indicates its structure. The "glut" indicates a five-carbon structure, which is similar to the five-carbon "straight chain" portion of citrate. Draw five carbons with a carboxylic acid group at each end and -CH_2 groups on carbons 2 and 4. The "3-hydroxy-3-methyl" portion of the molecule name indicates that carbon 3 has both a hydroxy group and a methyl group (which is carbon 6) attached. The "-yl" ending indicates a functional group joined to another functional group. Thus, pick one of the carboxylic acid groups and attach it in a thioester bond to coenzyme A. This reaction joins two molecules to form a larger molecule. The enzyme will then be either a *synthase* (no ATP used) or a *synthetase* (ATP used). This reaction does *not* use ATP, so the enzyme is a *synthase*. Synthases (and synthetases) are named for the *product*. Therefore the enzyme is called *HMG-CoA synthase*.

In reaction 3 (**figure 6.7**), a molecule of acetyl CoA is clipped off, which seems a bit wasteful because reaction 2 just added a third acetyl CoA. However, the formation of HMG-CoA is what drives the thiolase reaction (reaction 1) to run in reverse to make the acetoacetyl CoA. In reaction 3 the acetyl CoA molecule is clipped off by *HMG-CoA lyase* to produce the four-carbon product acetoacetate, which looks like acetoacetyl CoA, without the "CoA" attached. Note the carbon numbering on acetoacetate in **figure 6.7**. The methyl group (carbon 4) of acetoacetate was the methyl group (carbon 6) of HMG-CoA. The molecular name "acetoacetate" indicates its structure: two acetate molecules attached end to end (but not attached to CoA). *Acetoacetate* is the first ketone body formed.

Consider the structure of acetoacetate. It is a β-ketoacid, which means the carboxylic acid group can spontaneously decarboxylate to form CO_2. Again, "spontaneously" does not mean "instantaneously." As molecules of acetoacetate travel through the bloodstream, a proportion of them will spontaneously decarboxylate to form *acetone*, another ketone body. The structure of acetone was initially introduced in chapter 1 as part of the oxidation states flowchart (**figure 1.7**). Tissues cannot use acetone as a fuel source, because it cannot be converted back to two molecules of acetyl CoA. Thus, acetone is excreted from the body, either by breathing it off from the lungs or in the urine.

To allow more of the ketone bodies to be used as a fuel source rather than excreted, this pathway includes a fourth reaction (**figure 6.7**) that reduces the β-keto group of acetoacetate to a secondary alcohol, forming *β-hydroxybutyrate*. This ketone body, β-hydroxybutyrate, is four carbons, as indicated by the prefix "but" (which means "four" in organic chemistry nomenclature). The "-ate" ending indicates the deprotonated carboxylic acid group on an end carbon (carbon 1), with a hydroxy on the β-carbon (carbon 3). β-Hydroxybutyrate is the only ketone body that does *not* have a keto group.

The formation of β-hydroxybutyrate is a reduction reaction, so there must be a simultaneous oxidation reaction. Since the reduction from a ketone to an alcohol is going up the oxidation states flowchart (**figure 1.7**), one would expect NADPH + H^+ to be reduced to $NADP^+$. This reaction occurs in the mitochondrial matrix, though. NADPH is the form of reducing power used for synthetic reactions, which usually occur in the cytosol, and NADPH does not donate its electrons to the electron transport chain. Thus NADH is the predominantly available form of reducing power in the matrix, and therefore this reaction uses NADH + H$^+$ and oxidizes it to NAD$^+$ in this reaction. Enzymes that catalyze oxidation-reduction reactions using NAD$^+$ or FAD (regardless of which direction they are carried out) are called *dehydrogenases*. Dehydrogenases are named for the *more reduced* molecule, which is β-hydroxybutyrate in this reaction. The enzyme that catalyzes reaction 4 is called *β-hydroxybutyrate dehydrogenase*. This reaction is reversible. The β-hydroxybutyrate dehydrogenase enzymes of the tissues that take up ketone bodies do the reverse reaction to oxidize β-hydroxybutyrate back to acetoacetate, and simultaneously reduce NAD$^+$ to NADH + H$^+$. This oxidation of β-hydroxybutyrate would then "follow the rules," but note that the coenzyme used and the enzyme name is the same regardless of the direction of the reaction.

The amount of acetoacetate that is reduced to β-hydroxybutyrate by the liver is dependent on the ratio of NADH/NAD$^+$ in the matrix. The more NADH available, the more acetoacetate molecules are reduced to β-hydroxybutyrate. This reduced form, β-hydroxybutyrate, is the ketone body that is preferred for circulation in the blood because it cannot spontaneously decarboxylate like acetoacetate. Thus, more ketone bodies are available for use by the tissues. Of the three ketone bodies, β-hydroxybutyrate does not have a keto group. A particular clinical test uses sodium nitroprusside, which reacts with keto groups, to detect the amount of ketone bodies in a blood sample. This test, though, will only detect acetoacetate and acetone, the two ketone bodies containing keto groups. While the test will not detect the β-hydroxybutyrate, presumably the more prevalent ketone body in circulation in the blood, one can infer the relative amounts of ketone bodies based on the amounts of acetoacetate and acetone detected.

Conditions favorable for ketone body synthesis

A metabolic priority of the liver is to provide water-soluble fuel sources, in the form of glucose and ketone bodies, as readily available fuel sources (or fast food) for tissues to use. Glucose is preferred by the brain and is a necessity for mature red blood cells. Under conditions where the liver is maintaining blood glucose by de novo synthesis through gluconeogenesis, ketone body synthesis is also up-regulated. The brain can use ketone bodies as a fuel source, though usually after about forty-eight hours of starvation. Under starvation conditions, other tissues preferentially begin to utilize ketone bodies to reserve the available glucose for the brain and red blood cells.

The entry of acetyl CoA into the TCA cycle depends on the presence of oxaloacetate. Recall that oxaloacetate concentrations are kept low in the cell. Oxaloacetate is used almost as quickly as it is made. Oxaloacetate can be formed directly from pyruvate by pyruvate carboxylase (the first enzyme of gluconeogenesis) or from malate by malate dehydrogenase (the last step of the TCA cycle). When oxaloacetate is made it is used—either by citrate synthase of the TCA cycle or reduced by malate dehydrogenase during gluconeogenesis. In the case of ketone body synthesis, the low concentration of oxaloacetate in the matrix means that oxaloacetate is not available for citrate synthase. Under conditions when the cell is doing gluconeogenesis, oxaloacetate is formed from pyruvate carboxylase and citrate synthase is inhibited. The oxaloacetate is being reduced to malate, which is then transported to the cytosol for gluconeogenesis to continue. Under these conditions, pyruvate carboxylase is active and the PDH complex is inhibited. These are two enzymes in the same compartment that use the same substrate, pyruvate. The PDH complex and pyruvate carboxylase are regulated by high mitochondrial concentrations of acetyl CoA. The acetyl CoA used for ketone body synthesis comes from β-oxidation of fatty acids. As discussed, the catabolism of fatty acids yields many more acetyl CoA molecules than the catabolism of glucose.

Under what conditions (*when*), then, would oxaloacetate concentrations be low in the matrix (i.e., unavailable for use by the TCA cycle)? A low carbohydrate diet forces the liver to make glucose via gluconeogenesis to maintain blood glucose for other tissues, as would starvation conditions when a person is not eating anything. Uncontrolled diabetic conditions also result in the liver producing glucose and ketone bodies. There may be plenty of glucose in the blood, but without insulin the liver cannot detect the blood glucose levels. Therefore, the liver continues to secrete glucose and ketone bodies into the blood. Under these three conditions, gluconeogenesis predominates in the liver to form glucose because it is a metabolic priority.

Under these conditions, there is no oxaloacetate available in the matrix to run the TCA cycle. As gluconeogenesis is up-regulated in the liver, the catabolism of triacylglycerols from adipose cells is activated to provide fatty acids to the liver (via circulation in the blood attached to albumin). The fatty acids are transported into the mitochondrial matrix by the carnitine shuttle, where they are catabolized to acetyl CoA by β-oxidation. Since the oxaloacetate formed is now an intermediate of gluconeogenesis and not available for the citrate synthase reaction of the TCA cycle, the acetyl CoA produced by β-oxidation is used for ketone body synthesis.

In the liver the excess acetyl CoA formed by β-oxidation also simultaneously regulates two enzymes in the mitochondrial matrix that both use pyruvate as a substrate. A high concentration of acetyl CoA in the matrix inhibits the PDH complex, the bridging step between glycolysis and the TCA cycle. High concentrations of acetyl CoA in the matrix simultaneously activates pyruvate carboxylase, the first enzyme of gluconeogenesis. This simultaneous regulation of two enzymes that use the same substrate directs the use of pyruvate under conditions of excess acetyl CoA toward production of glucose via gluconeogenesis. The excess acetyl CoA from β-oxidation is then used for ketone body synthesis. Thus, the liver simultaneously activates gluconeogenesis and ketone body synthesis to provide the necessary water-soluble fuel sources to the blood for use by other tissues.

Fates of ketone bodies

The three ketone bodies are acetoacetate, β-hydroxybutyrate, and acetone. They are water soluble, which means they do not need a transporter for circulation in the blood. Both acetoacetate and β-hydroxybutyrate are four carbons. The tissues that take them up can convert them back to two molecules of acetyl CoA. The acetyl CoA derived from the ketone bodies can be used for the TCA cycle, fatty acid synthesis, or cholesterol synthesis—but not ketone body synthesis, which can only be done in the liver. Acetone, on the other hand, is only three carbons. Acetone is excreted. Acetone can be expired through the lungs, as it is volatile, and it can be excreted in the urine as well. Acetone is the molecule that produces the "fruity breath" smell from a person who is experiencing ketoacidosis due to uncontrolled diabetes. If ketone bodies are produced excessively, then all three ketone bodies can be excreted in the urine.

Tissues, such as skeletal muscle, that take up acetoacetate and β-hydroxybutyrate convert them back to two molecules of acetyl CoA. The ketone body β-hydroxybutyrate is first oxidized back to acetoacetate by *β-hydroxybutyrate dehydrogenase*, with simultaneous reduction of NAD^+ to $NADH + H^+$ (i.e., the reverse reaction of step 4 in **figure 6.7**). Molecules of acetoacetate are then attached to coenzyme A to form acetoacetyl CoA. The donor of coenzyme A is succinyl CoA, an intermediate of the TCA cycle. The removal of coenzyme A from succinyl CoA forms succinate, another intermediate of the TCA cycle. The enzyme that carries out this reaction is *β-ketoacyl CoA transferase*. This is the enzyme that liver cells lack, which prevents the liver from using ketone bodies as a fuel source. Now the molecules of acetoacetyl CoA can be cleaved to form two molecules of acetyl CoA by *β-ketothiolase*, also known as *acetoacetyl CoA thiolase* in this pathway. This is the same enzyme, though, that carries out this reaction during step 4 of the four repeated steps of β-oxidation of fatty acids.

When ketone body synthesis exceeds the rate at which they are used by the tissues, this excess ketone body production is called *ketosis*. The levels of ketone bodies begin to rise in the blood, referred to as *ketonemia*, and ultimately are excreted in excess in the urine, referred to as *ketonuria*. Excess ketone body production leads to what is called *ketoacidosis*. These molecules have carboxylic acids that deprotonate and contribute protons to the blood. Increasing proton concentration causes the blood pH to drop and become more acidic. Ketoacidosis is usually seen in uncontrolled type 1 diabetics, where ketone bodies are produced in excess and are not adequately taken up by tissues in relation to their production.

CHAPTER 7

FATTY ACID SYNTHESIS

OBJECTIVES

1. Define the process of fatty acid synthesis.
2. Explain the purpose of the pathway of fatty acid synthesis.
   a. Calculate how many cycles of the four repeated steps of fatty acid synthesis are necessary to generate the major product, palmitate ($C_{16:0}$ fatty acid).
   b. Explain, in general terms, how longer-chain fatty acids (C_{18} and higher) are synthesized and how double bonds are introduced into the acyl chain.
   c. Explain why linoleic and linolenic acids are dietary requirements.
3. Identify where fatty acid synthesis takes place in the cell.
   a. Explain how acetyl CoA, which is produced in the mitochondrial matrix, is shuttled to the cytosol and why a special export mechanism is required.
4. Explain how fatty acid synthesis is carried out in the cell.
   a. Identify the committed step of fatty acid synthesis, the enzyme that catalyzes this step, and its regulatory effectors.
   b. Describe the four main repeated steps of fatty acid synthesis and identify the enzymes and cofactor(s) involved.
   c. Explain the advantages of having all the enzymes involved in fatty acid synthesis (except acetyl CoA carboxylase) bound together in one large complex and the role the pantetheine group plays in the synthesis process.
5. Explain when fatty acid synthesis takes place.
   a. Identify the regulatory enzymes of fatty acid synthesis, what regulates them, and how hormones regulate fatty acid synthesis.
6. Compare and contrast the overall similarities and differences of the reciprocal pathways of β-oxidation of fatty acids and fatty acid synthesis.
   a. Compare and contrast the two pathways—including the four-repeated steps of each pathway, substrates, products, cofactors, enzymes, and regulation of the pathways.

FATTY ACID SYNTHESIS

The name of this pathway tells you its definition (*what*). Fatty acid synthesis is the synthesis of fatty acids from acetyl CoA. The typical fatty acid synthesized by the fatty acid synthase complex is palmitic acid (or palmitate, as the deprotonated form), a sixteen-carbon saturated fatty acid. Just like gluconeogenesis is not an exact reversal of glycolysis, neither is fatty acid synthesis an exact reversal of β-oxidation.

The fatty acids synthesized can be used for a variety of complex lipid syntheses (*why*). For instance, fatty acids are components of membrane lipids, triacylglycerols, and lipoproteins. Fatty acids, though, are not going to be synthesized to send into the mitochondria for immediate breakdown by β-oxidation, because that would be a futile cycle. A liver cell is not going to synthesize a fatty acid and then immediately use it for ketone body synthesis because β-oxidation precedes ketone body synthesis. So fatty acid synthesis and β-oxidation are reciprocal pathways that cannot occur in the same cell at the same time. There are distinct regulators that prevent these two pathways from being active at the same time in the same cell.

Fatty acid synthesis occurs in the cytosol of cells (*where*). The net reaction for fatty acid synthesis of palmitate is shown in **figure 7.1**. Fatty acid synthesis uses only one actual molecule of acetyl CoA along with seven molecules of malonyl CoA to synthesize a molecule of palmitate. As one would expect in a synthetic pathway, a lot of reducing power in the form of NADPH is required. However, ATP is not shown in this net reaction. Synthetic pathways should require an input of energy to make larger molecules from small starting units. The actual enzyme complex that condenses the two-carbon units together and does the necessary reductions and dehydration reaction does not require ATP. One could deduce this from the name of the enzyme complex—the *fatty acid synthase complex*. Synthases are enzymes that do not require energy. There is one enzyme involved in fatty acid synthesis, though, that is *not* part of the complex. *Acetyl CoA carboxylase* synthesizes malonyl CoA by carboxylation of acetyl CoA and requires ATP. The net reaction for malonyl CoA synthesis is also shown in **figure 7.1**. Only one acetyl CoA molecule is needed per fatty acid synthesized, and the remaining two-carbon units are attached using malonyl CoA. The synthesis of malonyl CoA must be done first to create a pool of these molecules for use by the fatty acid synthase complex, which is the energy requiring step for the fatty acid synthesis pathway.

This pathway is a bit more complicated than the catabolism of fatty acids (β-oxidation). One can think about this pathway as similar to an assembly line for car manufacturing. A car is built on an assembly line where parts of the car are added at each step before the entire car is released from the assembly line. The fatty acid synthase complex works the same way. All of the intermediates are added and reduced to the level of an alkane by the enzymes of the fatty acid synthase complex before the complete molecule of palmitate is released from the complex.

Palmitate synthesis	
1 acetyl CoA + 7 malonyl CoA + 14 NADPH + 20 H^+	⟶ 1 palmitate + 7 CO_2 + 8 CoA + 14 $NADP^+$ + 6 H_2O

Malonyl CoA synthesis	
7 acetyl CoA + 7 CO_2 + 7 ATP	⟶ 7 malonyl CoA + 7 ADP + 7 P_i + 14 H^+

Figure 7.1: The net reaction of palmitate (C_{16}) synthesis and malonyl CoA synthesis.

FIRST: Transport acetyl CoA from the mitochondrial matrix to the cytosol by way of citrate (using malate/citrate/pyruvate shuttle).
SECOND: Acetyl CoA is carboxylated to form malonyl CoA by the enzyme ***acetyl CoA carboxylase***.
THIRD: Chain growth by head-to-tail condensation and reduction to an alkane by the enzymes of the fatty acid synthase complex.

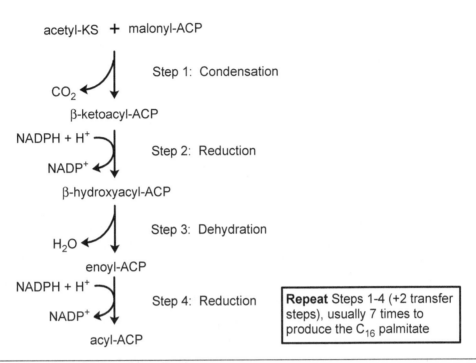

Figure 7.2: Overview of fatty acid synthesis.

The overview of the pathway of fatty acid synthesis is shown in **figure 7.2**. For learning the process (*how*) of fatty acid synthesis, one needs to know how molecules of acetyl CoA, which are generated in the mitochondrial matrix, are transported into the cytosol. A pool of malonyl CoA will then be produced from the acetyl CoA. For fatty acid synthesis there are six repeated steps that are carried out in the production of palmitate. Two of the steps are transfer steps to correctly position the growing fatty acid and a new molecule of malonyl CoA for continued lengthening of the carbon chain. There are four repeated steps that one should compare and contrast to the four repeated steps of β-oxidation of fatty acids, which will be referred to subsequently as the four key repeated steps. These four keys steps of focus are shown in the overview diagram for this pathway (**figure 7.2**). These four key repeated steps correspond to moving up the oxidation states flowchart (**figure 1.7**) to go from a more oxidized state of carbon to a more reduced state of carbon. In fact, exercise 5 at the end of the chapter 1 (the oxidation states chapter) required one to put in order the correct sequence of molecules for the last three reactions (of the four key repeated steps) of fatty acid synthesis based on their structures, as well as name the enzymes for these reactions.

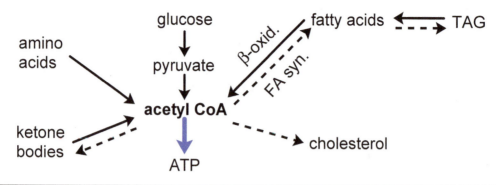

Figure 7.3: Sources and uses of acetyl CoA. (Note: The blue arrow indicates that acetyl CoA is used for ATP production via the TCA cycle and the reducing power that is sent to the electron transport chain.)

Figure 7.3 indicates the sources and uses of acetyl CoA. Excess nutrients in the diet (i.e., sugars and amino acids) can be converted to fatty acids for storage on triacylglycerols. Glycolysis and the PDH complex will yield a source of acetyl CoA that can be used for fatty acid synthesis. Tissues that take up ketone bodies could also use the acetyl CoA formed from them for fatty acid synthesis. The catabolism of some amino acids from protein breakdown also yield acetyl CoA, which could be used for fatty acid synthesis by the cell. Acetyl CoA can be used to make cholesterol, as well, but cholesterol cannot be catabolized back to acetyl CoA (as indicated by the one-way arrow in **figure 7.3**). Catabolism of fatty acids by β-oxidation also does not provide the pool of acetyl CoA for fatty acid synthesis. Recall that β-oxidation and fatty acid synthesis are reciprocal pathways and will not be active at the same time in the same cell. Excess fatty acids obtained from the diet can be used for the production of body stores of triacylglycerols or in membrane lipid synthesis. A primary function of the liver is to store all the excess nutrients obtained from the diet. The liver only stores a certain amount of glycogen, as a readily accessible source of glucose. The remaining excess nutrients are stored as triacylglycerols, the structure of which is three fatty acids attached to a glycerol backbone.

Triacylglycerols (TAGs or fat) are the most efficient storage form of fuel, as discussed previously, and are not rapidly mobilized. Triacylglycerols can be stored for 150 days or longer. Glycogen is the storage form of glucose. For most tissues that store glycogen, it is for the cell's own use. Glycogen supplies a ready source of glucose, particularly for catabolism to produce ATP. As the liver is the primary organ responsible for maintaining blood glucose, its glycogen stores are the only glycogen stores that can release glucose into the blood for use by all other tissues. However, an adult liver can only store about a twenty-four hour supply of glucose. Proteins in the body are constantly turned over (i.e., enzyme turnover), but proteins are not an energy source that is regularly used by cells. There is no protein whose sole function is to serve as an energy store for the cell like glycogen and triacylglycerols. When a protein is broken down, along with the further catabolism of its amino acids, it does yield ATP. However, the actual function of that protein is lost, whether its function was as a transporter, enzymatic, structural (i.e., muscle), or some other function. Some of the longer-functioning proteins are turned over in a maximum of about ten days.

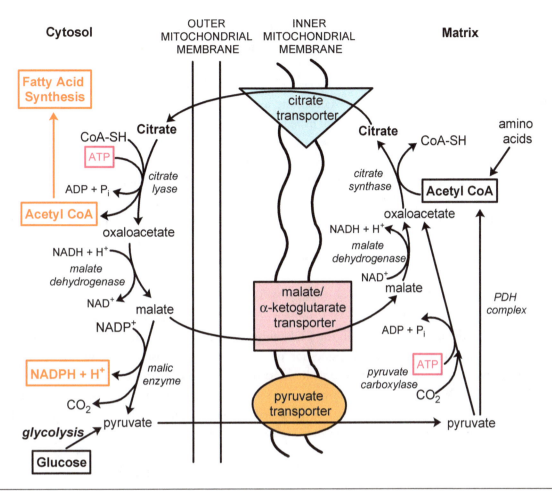

Figure 7.4: Malate/citrate/pyruvate shuttle.

For the fatty acid synthesis pathway, there is a transport issue that must first be discussed. The acetyl CoA produced for this pathway (i.e., from the PDH complex, amino acid catabolism, or the use of ketone bodies) is in the mitochondrial matrix. Fatty acid synthesis takes place in the cytosol, so the acetyl CoA must be transported from the matrix to the cytosol. **Figure 7.4** shows the malate/citrate/pyruvate shuttle that is used to transport acetyl CoA from the matrix to the cytosol. In **figure 7.4**, start with the acetyl CoA in the *matrix*, which comes from glucose (via glycolysis and the PDH complex) or amino acid catabolism. Under conditions where the cell needs to synthesize fatty acids, *citrate synthase* condenses acetyl CoA with oxaloacetate to produce citrate, which is the first reaction of the TCA cycle (**figure 3.7**). Rather than continuing through the TCA cycle, citrate is now transported across the inner mitochondrial membrane (through a specific transporter) to the cytosol. Thus, molecules of citrate serve as the carriers of the acetyl units to the cytosol. Once the citrate is in the cytosol, an enzyme called *citrate lyase* cleaves citrate to re-form acetyl CoA and oxaloacetate. Now a pool of acetyl CoA is available to be used for fatty acid synthesis or cholesterol synthesis, as both pathways take place in the cytosol of a cell.

As the TCA cycle is not running in the matrix under these conditions, the oxaloacetate formed in the cytosol needs to be transported back into the matrix (**figure 7.4**). However, as previously discussed, oxaloacetate does not have a transporter in the inner mitochondrial membrane. Malate is allowed to cross, so the oxaloacetate can be reduced to malate by cytosolic malate dehydrogenase. Malate is then transported back to the matrix to be re-oxidized to oxaloacetate by mitochondrial malate dehydrogenase. Fatty acid synthesis requires a lot of NADPH, the form of reducing power needed to donate electrons for the reduction of molecules in synthetic processes. The majority of NADPH is produced by the pentose phosphate pathway. However, additional NADPH can be formed from another enzymatic reaction in this shuttle. A cytosolic enzyme called *malic enzyme* (which does not follow any of the naming rules, including the "-ase" ending used for enzymes) produces pyruvate from the decarboxylation and oxidation of malate (shown in **figure 7.4**). The purpose of this step produces additional molecules of NADPH needed for fatty acid synthesis. The ratio of the amount of malate versus pyruvate transported back to the matrix will depend on the ratio of $NADPH/NADP^+$ in the cytosol needed for fatty acid synthesis.

The pyruvate produced by malic enzyme can also be transported back into the matrix where it can be used to form more acetyl CoA (via the PDH complex) or possibly oxaloacetate (via pyruvate carboxylase). The possibility of the matrix having both the PDH complex and pyruvate carboxylase active under these conditions is a complicating factor to the question of *when* for one to consider. One would not generally expect two enzymes that use the same substrate in the same cellular compartment, as in this case, to both be on at the same time. What happens to the pyruvate in the matrix is dependent on the need for oxaloacetate based on the ratio of malate versus pyruvate being shuttled back into the matrix, as well as the amount of acetyl CoA that needs to be sent to the cytosol for fatty acid synthesis.

The cell now has a pool of acetyl CoA in the cytosol. The first reaction of fatty acid synthesis, shown in **figure 7.5**, is *not* part of the fatty acid synthase complex. This enzymatic reaction must first create a pool of malonyl CoA molecules for use by the complex. This reaction is the carboxylation of acetyl CoA to form malonyl CoA, which requires energy in the form of ATP. The enzyme is a *carboxylase* and is named for the molecule accepting the CO_2, acetyl CoA. Therefore, the enzyme that catalyzes this reaction is called *acetyl CoA carboxylase*.

Recall that carboxylase enzymes require the coenzyme biotin. The reaction catalyzed by acetyl CoA carboxylase is analogous to the two-step reaction catalyzed by pyruvate carboxylase (**figure 5.5**) in gluconeogenesis.

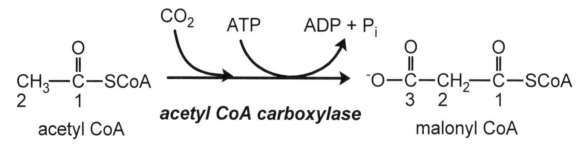

Figure 7.5: Formation of malonyl CoA, the committed step in fatty acid synthesis.

Acetyl CoA carboxylase first attaches the CO_2 group (from bicarbonate, HCO_3^-) to its biotin coenzyme, which requires ATP. The carboxylic acid group is then transferred to the methyl carbon (carbon 2) of acetyl CoA to form malonyl CoA. Malonyl CoA is now three carbons containing a carboxylic acid group at each end. The carboxylic acid carbon that is designated carbon 1 of malonyl CoA, though, is the one that is joined to coenzyme A via a thioester bond (as shown in **figure 7.5**).

The acetyl CoA carboxylase reaction is *irreversible* and the committed step of fatty acid synthesis. Thus, this is the key regulated step of the pathway. The removal of the carboxylic acid group (carbon 3) will drive the condensation reaction, the first reaction of the four key repeated steps of fatty acid synthesis.

The fatty acid synthase complex is very large. In humans (and other vertebrates), the fatty acid synthase complex is a dimer of two identical subunits that lie head-to-tail relative to one another. Each subunit of the dimer is composed of a single large polypeptide containing all seven enzymes (i.e., two transfer step enzymes, the four enzymes for the four key repeated catalytic steps, as well as the enzyme that clips off the final product). See **table 7.1** for a list of all seven enzymes and the reactions they carry out. Note that the name of the complex is a *synthase*, which are enzymes that do not use energy in the form of ATP. The synthase is named for the *product* of the complex—a fatty acid. Thus, the complex is called the *fatty acid synthase complex*.

The fatty acid synthase complex also has an acyl carrier protein, abbreviated ACP. The ACP is derived from pantothenic acid (like coenzyme A). The ACP has a free sulfhydryl group (–SH) at its terminus. Acyl groups

Table 7.1: The enzymatic components of the fatty acid synthase complex

STEP	COMPONENT	REACTION/FUNCTION	REPEATED STEP
Transfer step 1	Acetyl CoA-ACP transacylase	Transfers first acetyl CoA or growing acyl chains to the cysteine of the β-ketoacyl-ACP synthase	Yes
Transfer step 2	Malonyl CoA-ACP transacylase	Transfers malonyl units onto the sulfhydryl group of the ACP	Yes
Key reaction step 3	β-ketoacyl-ACP synthase (a.k.a. the condensing enzyme)	Condenses the acetyl unit or growing acyl chain onto carbon 2 of the malonyl units on the ACP, with loss of CO_2	Yes
Key reaction step 4	β-ketoacyl-ACP reductase	Reduces the keto group on the β-carbon to a hydroxy group, with oxidation of NADPH	Yes
Key reaction step 5	β-hydroxyacyl-ACP dehydratase	Dehydrates to form a *trans*-alkene between carbons 2 and 3	Yes
Key reaction step 6	Enoyl-ACP reductase	Reduces the alkene to an alkane, with oxidation of NADPH	Yes
Final reaction step 7	palmitoyl thioesterase	Clips off palmitate, the final product	No

are attached to the ACP via a thioester bond, in the same manner that fatty acids are attached to coenzyme A molecules. The ACP serves as a "long arm" that can reach all of the enzyme active sites of the complex. The function of the ACP, then, is to carry the acyl intermediates around to each enzyme of the complex for the subsequent rounds of the four key repeated steps of fatty acid synthesis.

There are two transfer steps that are repeated (see **table 7.1**). These two transfer steps are to place the initial acetyl CoA for the first round of repeated steps (or the growing acyl chain in the following rounds) onto a cysteine group of the β-ketoacyl-ACP synthase. This first transfer step ensures that the malonyl unit is always transferred onto the sulfhydryl group of the ACP, which is the second transfer step. The condensation of the growing acyl chain onto the new malonyl unit allows for the ACP to carry it to the other three enzyme active sites of the complex to reduce the newest β-keto group. The four key repeated steps of fatty acid synthesis that coincide with going up the oxidation states flowchart (**figure 1.7**) are the focus and are shown in **figure 7.6**.

Figure 7.6 shows the four key repeated steps of fatty acid synthesis. These are the four steps one should be able to compare and contrast with the pathway of β-oxidation. The reactions of the fatty acid synthase complex begin with the transfer steps, which are *not* shown in **figure 7.6**. *Acetyl CoA-ACP transacylase* transfers the two-carbon unit from a molecule of acetyl CoA to the -SH group of a cysteine residue of the β-ketoacyl-ACP synthase, abbreviated KS. The acetyl unit is shown attached to KS as a starting material of step 1 in **figure 7.6**. The three-carbon unit from a molecule of malonyl CoA is then transferred to the -SH group of the ACP by the enzyme *malonyl CoA-ACP transacylase*. In **figure 7.6** the malonyl unit is shown attached to the ACP as the other starting material of step 1.

In the pathway of β-oxidation, the β-carbon needed to be oxidized to a keto group. Therefore, the four repeated steps were oxidation to an alkene, hydration to a secondary alcohol, oxidation to a keto group, then cleave off an acetyl CoA. In fatty acid synthesis, two two-carbon units need to be connected together first to create a four-carbon molecule. The first key reaction, as shown in **figure 7.6**, is a condensation reaction—the attachment of the two two-carbon units together. In this reaction, the decarboxylation (loss of CO_2) of the malonyl unit (carbon 3) drives the condensation of the acetyl unit *onto* carbon 2 of the malonyl unit. (In subsequent rounds, it will be the growing acyl chain that condenses onto carbon 2 of the new malonyl unit.) The product of this first condensation reaction is a four-carbon intermediate attached to the ACP, called β-ketoacyl-ACP.

This first key reaction involves attaching two molecules together to make a larger molecule. Recall that enzymes that catalyze synthesis reactions are called *synthases* or *synthetases*, depending on whether or not ATP is used, and are named for the *product* of the reaction. As this reaction does not require ATP, the enzyme is called a *synthase*, and the product is β-ketoacyl-ACP. Therefore, the enzyme for reaction 1 is called β-ketoacyl-ACP synthase (KS). The "ACP" as part of the enzyme names of the fatty acid synthase complex should serve as one way to easily distinguish the enzyme names of the four repeated steps of fatty acid synthesis versus β-oxidation. The β-ketoacyl-ACP synthase enzyme is often referred to as the *condensing enzyme*.

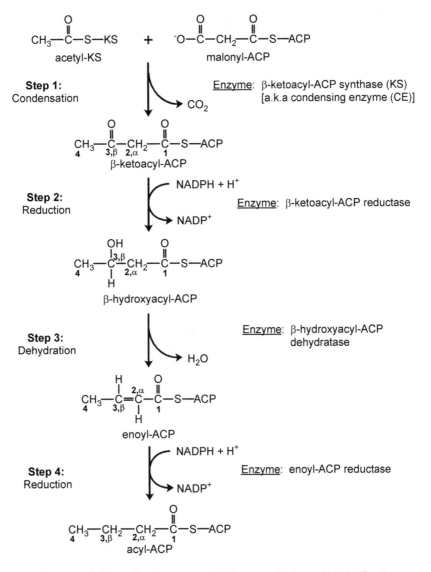

Figure 7.6: The four key repeated steps of the fatty acid synthase complex. (Note: There are actually six repeated steps for each two-carbon addition. The other two steps are "transfer" steps: first, to move the growing acyl chain back to the sulfhydryl group of the KS; second, to attach the next malonyl group to the ACP. Also note that all subsequent two-carbon additions come from *malonyl CoA*. The only acetyl CoA used is in the first round.)

This four-carbon product of reaction 1 looks like a β-ketoacyl CoA intermediate of β-oxidation (i.e., the product of reaction 3 in β-oxidation), only now the four-carbon unit is attached to the ACP instead of coenzyme A. Recall the purpose of this pathway is to make a sixteen-carbon saturated fatty acid. Fatty acids have only one carboxylic acid group at carbon 1. The remaining carbons are all at the level of an alkane for a saturated fatty acid. The structure of the product, β-ketoacyl-ACP, has a keto group on carbon 3 (the β-carbon). This keto group needs to be reduced back to the level of an alkane. According to the oxidation states flowchart (**figure 1.7**), the reaction sequence needed to accomplish this is to reduce the keto group to a secondary alcohol, dehydrate to form an alkene, then reduce the alkene to an alkane. The next three enzymes of the fatty acid synthase complex carry out these exact three reactions.

The second key reaction, shown in **figure 7.6**, is the reduction of the β-keto group (carbon 3) on the β-ketoacyl-ACP to a secondary alcohol, forming β-hydroxyacyl-ACP. The reduction reaction requires a simultaneous oxidation reaction. When going up the oxidation states flowchart, NADPH + H$^+$ is the form of reducing power used, and it will be oxidized to NADP$^+$. Enzymes that catalyze oxidation-reduction reactions using NADP$^+$ or NADPH, regardless of which direction the reaction is going, are called *reductases*. Reductases are named for the *more oxidized* molecule because that is what is being reduced. In this reaction, the β-ketoacyl-ACP molecule is more oxidized, so the enzyme is called *β-ketoacyl-ACP reductase*.

In step 3 (**figure 7.6**), a molecule of water is removed (i.e., dehydration) from the β-hydroxyacyl-ACP to form an alkene between carbons 2 and 3 in the product, enoyl-ACP. This is also a *trans* double bond, like in β-oxidation, and will be removed in the following step. The enzyme that catalyzes this dehydration reaction is called a *dehydratase*, and it is named for the molecule that is losing the water—β-hydroxyacyl-ACP. The enzyme is called *β-hydroxyacyl-ACP dehydratase*.

Step 4 (**figure 7.6**) now reduces the alkene (enoyl-ACP) to an alkane, forming an acyl-ACP. For this reduction the form of reducing power is still NADPH + H$^+$, which will be oxidized to NADP$^+$. An enzyme that catalyzes an oxidation-reduction reaction using NADP$^+$/NADPH, regardless of the direction of the reaction, is called a *reductase*. A reductase is named for the *more oxidized* molecule, which is enoyl-ACP for this reaction. This enzyme is called *enoyl-ACP reductase*. The product of these four key repeated steps is a four-carbon fatty acid attached to the ACP of the complex.

Before another two-carbon unit can be added, the acetyl CoA-ACP transacylase transfers this four-carbon acyl unit to the –SH group on the cysteine residue of the β-ketoacyl-ACP synthase enzyme (KS). Then the malonyl CoA-ACP transacylase transfers a three-carbon unit from a new malonyl CoA molecule and attaches it to the -SH group of the ACP. Note that for all subsequent rounds of fatty acid synthesis, malonyl CoA is the donor of the two-carbon units. The only actual molecule of acetyl CoA used is in the first round. The β-ketoacyl-ACP synthase then condenses the four-carbon acyl unit onto carbon 2 of the new malonyl unit, with the loss of CO_2. Now a six-carbon unit has been formed that has a β-keto group. This β-keto group will be reduced back to an alkane by the sequence of reduction, dehydration, and reduction reactions. In subsequent rounds of fatty acid synthesis, the acetyl CoA-ACP transacylase will move the growing acyl chain (after step 4) from the ACP to the cysteine residue of the β-ketoacyl-ACP synthase. This allows the new malonyl unit to always be transferred to the ACP by malonyl CoA-ACP transacylase. Note that the names of these transfer enzymes indicate the reactions they carry out.

It requires seven cycles of these four repeated steps (six repeated steps when including the two transfer steps) to produce the typical product palmitate, a sixteen-carbon saturated fatty acid. When the sixteen-carbon fatty acid has been formed, the enzyme *palmitoyl thioesterase* cleaves the thioester bond that attaches the acyl group to the ACP, forming palmitate. Palmitate will then be attached to coenzyme A, forming palmitoyl CoA, by the enzyme fatty acyl CoA synthetase (**figure 6.3**). Fatty acids must always be attached to a carrier, as previously discussed.

Further processing of newly synthesized fatty acids

Palmitate is not the only fatty acid in the body, though it is the major product of the fatty acid synthase complex. Further processing of fatty acids occurs in the endoplasmic reticulum. Longer fatty acids can be produced by elongating palmitate (or other fatty acids) using molecules of malonyl CoA as the two-carbon donor, similar to the fatty acid synthase complex. These fatty acids can be eighteen carbons, twenty carbons, or longer. Malonyl CoA is used to sequentially add two carbons to the *carboxyl end* of both saturated and unsaturated fatty acids.

Desaturation reactions can occur to introduce one or more *cis* double bonds (i.e., alkenes) to form unsaturated fatty acids. Recall that *cis* double bonds produce a bend, or "kink," in the acyl chain of a fatty acid, preventing tight packing of fatty acids—especially in the membranes. The reactions to make double bonds are carried out by several enzymes (termed mixed-function oxidases), including a reductase and a desaturase. One or more *cis* double bonds can be introduced at various positions along the acyl chain.

Humans, however, cannot introduce double bonds past carbon 9 of a fatty acid. Palmitate, the sixteen-carbon saturated fatty acid, is the primary product of the fatty acid synthase complex. Without the capability of introducing double bonds past carbon 9, people must obtain long-chain fatty acids with double bonds already positioned past carbon 9 in their diet in order to synthesize some important molecules. Linoleic (ω-6) and linolenic (ω-3) fatty acids are essential fatty acids that must be obtained in our diet. Linoleic acid ($C_{18:2}^{\Delta 9,12}$) is an eighteen-carbon fatty acid with double bonds at carbon 9 and 12 (i.e., between carbons 9 and 10; and between carbons 12 and 13). Linolenic acid ($C_{18:3}^{\Delta 9,12,15}$) is an eighteen-carbon fatty acid with three double bonds at carbons 9, 12, and 15. The omega-3 fatty acids, like linolenic acid, are typically found in fish oils. The omega-6 fatty acids, like linoleic acid, are usually found in seed oils (i.e., vegetable oils). By obtaining these essential fatty acids in the diet, the cells can further modify them to produce other important fatty acids. Arachidonate ($C_{20:4}^{\Delta 5,8,11,14}$) is a twenty-carbon fatty acid with four double bonds synthesized from linoleic acid. Eicosapentaenoic acid (EPA), $C_{20:5}^{\Delta 5,8,11,14,17}$, is derived from linolenic acid. Both arachidonate and EPA are stored in membranes as part of various membrane phospholipids and help regulate membrane fluidity. The release of arachidonate or EPA leads to the production of different sets of eicosanoid hormones (i.e., prostaglandins, leukotrienes, and thromboxanes). The various eicosanoid hormones mediate numerous physiological processes such as smooth muscle contraction, regulation of blood flow, inflammation, fever, pain perception, and platelet aggregation.

Regulation of fatty acid synthesis

Regulation of the fatty acid synthesis pathway begins with the committed step, catalyzed by acetyl CoA carboxylase. Recall that acetyl CoA carboxylase is not part of the fatty acid synthase complex. It is regulated by hormones such as insulin and glucagon. Insulin will activate the carboxylase, as an indicator of the fed state, allowing excess nutrients to be stored as triacylglycerols. Glucagon inhibits the acetyl CoA carboxylase. Under these conditions, the liver (as a primary target tissue of glucagon) would be doing gluconeogenesis and ketone body synthesis to release water-soluble fuel sources into the blood. The acetyl CoA for ketone body synthesis comes from β-oxidation under these conditions. Fatty acid synthesis would be inhibited in a cell that is doing β-oxidation.

Citrate is an allosteric activator of acetyl CoA carboxylase. As the carrier of acetyl units to the cytosol from the matrix, citrate is a cellular indicator that ATP and building blocks are abundant for fatty acid synthesis to occur. Palmitoyl CoA inhibits the acetyl CoA carboxylase. If palmitoyl CoA is abundant in the cytosol, it generally means that either fatty acids are being cut off of lipids (i.e., triacylglycerols or phospholipids) or are being brought into the cell to serve as a fuel source. Under these conditions, these fatty acyl CoA molecules are being transported into the matrix for β-oxidation to occur. Thus fatty acid synthesis would be inhibited at the acetyl CoA carboxylase reaction. AMP also serves as an inhibitor of acetyl CoA carboxylase because high levels of AMP mean ATP levels are low. ATP is needed for synthesis of fatty acids.

Table 7.2 summarizes the reciprocal regulation of the two pathways of β-oxidation and fatty acid synthesis. Palmitoyl CoA, as previously mentioned, is an inhibitor fatty acid synthesis by inhibiting acetyl CoA carboxylase. Malonyl CoA, the main substrate for fatty acid synthesis, is an inhibitor of β-oxidation. Malonyl CoA specifically inhibits carnitine acyltransferase I (CAT I).

Insulin, as a hormonal signal of the fed state, results in the dephosphorylation of key metabolic enzymes of target tissues, usually to activate them. Glucagon, as a hormonal indicator of low blood glucose, leads to the phosphorylation of key metabolic enzymes in its target tissues, usually to inhibit them. Under insulin conditions, the cells can take up nutrients and do synthetic processes, like fatty acid synthesis, because fuel sources are readily abundant. For fatty acid synthesis, insulin dephosphorylates acetyl CoA carboxylase to

Table 7.2: Reciprocal regulation of β-oxidation of fatty acids and fatty acid synthesis

	SUBSTRATE	REGULATED ENZYME	REGULATORS
β-oxidation	Palmitoyl CoA	Carnitine acyl transferase I (CAT I)	Malonyl CoA (−) Insulin (−) Glucagon (+)
Fatty acid synthesis	Malonyl CoA	Acetyl CoA carboxylase	Palmitoyl CoA (−) Insulin (+) Glucagon (−)

activate it. The increase in concentration of malonyl CoA in the cytosol by acetyl CoA carboxylase inhibits CAT I, preventing the newly synthesized fatty acids from entering the mitochondrial matrix, and as a result, β-oxidation is inhibited.

When glucagon is the predominant hormone, the liver needs to release water-soluble fuel sources into the blood in the form of glucose and ketone bodies. Glucagon leads to the phosphorylation of acetyl CoA carboxylase, inhibiting it and fatty acid synthesis overall. The decrease in malonyl CoA in the cytosol relieves the inhibition of CAT I, allowing the influx of fatty acids to enter the mitochondrial matrix for β-oxidation and ultimately ketone body synthesis.

COMPARISON OF β–OXIDATION AND FATTY ACID SYNTHESIS

Table 7.3 and **figure 7.7** show a couple of ways to compare and contrast the pathways of β-oxidation and fatty acid synthesis. **Table 7.3** summarizes key distinguishing features of the two pathways, such as cellular locations of the pathways, carriers of the intermediates, and forms of reducing power formed or used.

Table 7.3: Features distinguishing β-oxidation and fatty acid synthesis

FEATURE	ß-OXIDATION	FATTY ACID SYNTHESIS
Location of pathway in cell	Mitochondria	Cytosol
Carrier of intermediate acyl groups	CoA-SH	ACP-SH
Organization of enzymes	None	Fatty acid synthase complex
Form in which two-carbon units participate	Acetyl CoA	Malonyl CoA
Electron donor or acceptor used	FAD, NAD$^+$	NADPH
Participation of CO_2	No	Yes
Stereochemistry of β-hydroxyacyl intermediate	L	D

Figure 7.7 compares the four repeated steps of β-oxidation of fatty acids with the four key repeated steps of fatty acid synthesis. On the right side of **figure 7.7** is β-oxidation, while the left side shows fatty acid synthesis. The organization of the reactions is analogous to the oxidation states flowchart (**figure 1.7**) in that the molecules become more oxidized as one moves down the page. As one moves up the page, the molecules are more reduced. The reactions on the right for β-oxidation go down the page and are (1) dehydrogenation (a.k.a. oxidation), (2) hydration, (3) dehydrogenation (a.k.a. oxidation), and (4) thiolytic cleavage. On the left, start at the bottom and go up to carry out (1) condensation, (2) reduction, (3) dehydration, and

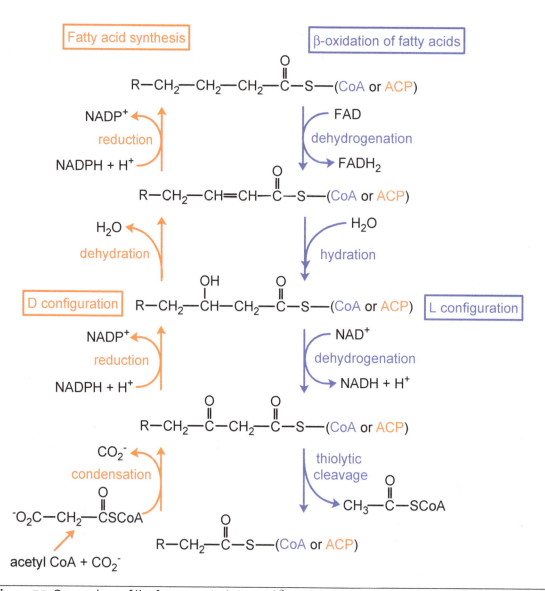

Figure 7.7: Comparison of the four repeated steps of β-oxidation and fatty acid synthesis pathways.

(4) reduction. Also pay attention to whether the carrier is coenzyme A (β-oxidation) or the ACP for fatty acid synthesis.

INTEGRATION OF CARBOHYDRATE AND LIPID METABOLISM

The last two figures, **figure 7.8** and **figure 7.9**, illustrate how carbohydrate and lipid pathways are integrated under specific conditions by liver cells, as an example. These figures further address the question of *when* for metabolic pathways. In both figures, the heavy, solid arrows indicate the pathways occurring under the stated conditions, while the dotted lines indicate points of regulation.

An important function of the liver is to sense the nutrient needs of the body and supply nutrients via the bloodstream, especially water-soluble fuel sources. The liver also serves to store excess nutrients, which can be used to replenish nutrients in the bloodstream as other tissues use them. Nutrients from the diet are absorbed and sent to the liver via portal blood circulation. If blood levels of particular nutrients are low, the liver filters the blood to remove toxins and foreign particles but does not take up the nutrients. Glucose, as an example, is allowed to pass into systemic blood circulation until the levels of glucose reach about 5 mM. As blood glucose levels rise above 5 mM, liver cells begin to take up glucose. As discussed in chapter 2, the key differences in the isozymes of glucokinase (in liver cells) versus hexokinase (in other tissues) allow for the liver to take up glucose and phosphorylate it only when the blood levels are rising above a particular concentration range. In between meals the liver functions to ensure that appropriate nutrients are available in the bloodstream for use by other tissues. In particular the liver must secrete the water-soluble fuel sources, glucose and ketone bodies, so that tissues always have ready access to a quick source of fuel.

In the well-fed state (i.e., after eating a meal), shown in **figure 7.8**, the insulin to glucagon ratio in the blood would rise, indicating more insulin is being released by the pancreas. Under these conditions, the liver (a primary target tissue of insulin) would first allow nutrients to pass into systemic circulation (if necessary) to ensure other tissues have their nutrient needs met. If nutrients are still coming in from having eaten surplus calories (i.e., excess glucose, fatty acids, and amino acids), liver cells will begin to take up these additional nutrients and use them or store them. In **figure 7.8**, excess glucose (upper left of diagram) will first be used to replenish liver glycogen stores. Once the liver glycogen stores are sufficient, the additional glucose will be stored as triacylglycerols. In **figure 7.8** follow the heavy dark arrows. Start with the excess glucose and follow the arrows through glycolysis to pyruvate, which moves into the mitochondrial matrix. The pyruvate is then converted to acetyl CoA by the PDH complex, which is active under these conditions. The acetyl CoA combines with oxaloacetate to form citrate using citrate synthase. The citrate, though, is exported to the cytosol (rather than continuing through the TCA cycle). Once in the cytosol, the citrate is cleaved to re-form acetyl CoA and oxaloacetate. The oxaloacetate goes back to the mitochondrial matrix via the malate/citrate/pyruvate shuttle (as shown in detail in **figure 7.4**). The cytosolic acetyl CoA can be used for acyl CoA synthesis, which are stored as part of *triacylglycerols*, or for *cholesterol* synthesis. The malonyl CoA formed in the process of fatty acid synthesis inhibits CAT I, which prevents the newly formed acyl CoA molecules from entering the mitochondria for breakdown by β-oxidation. The newly synthesized fatty acids will be attached to glycerol backbones to form triacylglycerols, the most efficient form of fuel storage for the body.

The breakdown of excess amino acids yields products including pyruvate, acetyl CoA, and various TCA cycle intermediates. Thus, excess intake of amino acids can also be used for the synthesis of glucose (for storage as glycogen) or fatty acids (for storage as triacylglycerols). Excess dietary lipids are not shown **figure 7.8**, as they would be circulated as part of triacylglycerols in chylomicrons (a type of lipoprotein particle) for use by other tissues, especially adipose tissue. Adipocytes (adipose cells) take up the fatty acids from chylomicrons for storage as triacylglycerols, which is a primary function of adipose tissue. In summary, liver cells under

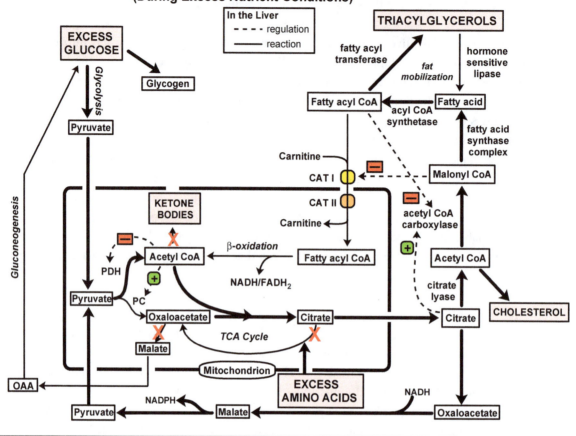

Figure 7.8: The integration of carbohydrate and lipid metabolism in the liver in the well-fed state.

excess nutrient conditions replenish their glycogen stores and convert additional surplus nutrients to fatty acids for storage as triacylglycerols.

In the starved state (or insulin deficiency state), the insulin to glucagon ratio drops indicating that glucagon is the predominant hormone in the bloodstream. Note that individuals with type 1 diabetes are unable to produce insulin. The liver is also a primary target tissue of glucagon. Under these conditions, shown in **figure 7.9**, the liver responds by up-regulating pathways that will supply water-soluble fuel sources to the blood. These two water-soluble fuel sources are glucose and ketone bodies. In the case of starvation beyond twenty-four hours or in uncontrolled diabetes (lack of insulin), the glycogen stores are used up. As indicated in **figure 7.9**, the liver has up-regulated gluconeogenesis and ketone body synthesis to provide the necessary circulating fuel sources for the body. The sources of pyruvate for gluconeogenesis primarily come from lactate and amino acids (from protein breakdown). In **figure 7.9** follow the heavy dark arrow from pyruvate (in the mitochondrial matrix) to oxaloacetate and then to malate. The malate is exported to the cytosol and re-oxidized to oxaloacetate, which continues through gluconeogenesis to produce glucose. The liver cells secrete the glucose into the blood (not shown in the figure).